Hydrology, Hydraulics, and Geomorphology of the Bonneville Flood

Jim E. O'Connor*
Department of Geosciences
University of Arizona
Tucson, Arizona 85721

THE GEOLOGICAL SOCIETY OF AMERICA · 1888 ·

SPECIAL PAPER
274
1993

*Present address: U.S. Geological Survey, 5400 MacArthur Boulevard, Vancouver, Washington 98661.

Published by The Geological Society of America, Inc.
3300 Penrose Place, P.O. Box 9140, Boulder, Colorado 80301

Printed in U.S.A.

GSA Books Science Editor Richard A. Hoppin

Library of Congress Cataloging-in-Publication Data
O'Connor, Jim E., 1959–
Hydrology, hydraulics, and geomorphology of the Bonneville flood / Jim E. O'Connor.
p. cm. — (Special paper / Geological Society of America ; 274)
Includes bibliographical references.
ISBN 0-8137-2274-8
1. Geology, Stratigraphic—Pleistocene. 2. Floods—Idaho—Snake River Valley. 3. Geology—Idaho—Snake River Valley. 4. Geomorphology—Idaho—Snake River Valley. I. Title. II. Series: Special paper (Geological Society of America) ; 274.
QE696.023 1993
551.48'9'097961—dc20 92-42641
CIP

Front cover photo: Oblique aerial view to the east (upstream) of the Snake River Plain and the Snake River Canyon near Murtaugh, Idaho. The maximum extent of erosion by the Bonneville Flood is shown by the limits of cultivation on the uplands. The large set of falls in the middle foreground is Cauldron Lynn. At the peak discharge, only about a third of the flow followed the Snake River Canyon in this reach; the remainder flowed over an upland route to the north and reentered the Snake River Canyon near Twin Falls. **Back cover photo:** Large and rounded boulders deposited by the Bonneville Flood within the Snake River Canyon near Swan Falls, Idaho. Long diameters of several of the boulders exceed ten meters. The view is to the southwest.

10 9 8 7 6 5 4 3 2 1

Dedication

Keith Katzer
June 14, 1960–July 8, 1988

The joy in completing this project has been tempered by the absence of Keith Katzer. Keith was a major factor in the development of my geologic career as well as my life. A part of me has gone with him.

Contents

Geological Society of America
Special Paper 274
1993

Hydrology, Hydraulics, and Geomorphology of the Bonneville Flood

ABSTRACT

Approximately 14,500 yr ago, Pleistocene Lake Bonneville discharged 4,750 km^3 of water over the divide between the closed Bonneville basin and the watershed of the Snake River. The resulting flood, released near Red Rock Pass, Idaho, followed the present courses of Marsh Creek, the Portneuf River, and the Snake and Columbia Rivers before reaching the Pacific Ocean. For 1,100 km between Red Rock Pass and Lewiston, Idaho, the Bonneville Flood left a spectacular array of flood features that have allowed for geologic reconstruction and quantitative evaluation of many aspects of the flood hydrology, hydraulics, erosion, and sediment transport.

Geologic evidence of maximum flood stages in conjunction with step-backwater flow modeling has furnished estimates of peak discharge and local hydraulic flow conditions for 10 separate reaches along the flood route. Peak discharge was approximately 1.0 million m^3·sec^{-1} at the Lake Bonneville outlet near Red Rock Pass. Downstream, near Lewiston, Idaho, the maximum discharge attenuated to 0.57 to 0.62 million m^3·sec^{-1}. In part, attenuation of the peak discharge was probably the result of flow storage in the wide alluvial valleys of the western Snake River Plain.

Diverse geologic and geomorphic environments along the flood route resulted in large spatial variations in the channel geometry as the flood moved downstream. Consequently, the local hydraulic conditions (flow depth, velocity, boundary shear stress) of the Bonneville Flood varied substantially within and between the study reaches. The rate of energy expenditure was also highly varied; local calculated stream power magnitudes ranged from less than 10 watts·m^{-2} in ponded reaches to more than 100,000 watts·m^{-2} in major constrictions. More than 50% of the total energy loss at peak discharge was expended over a distance that encompassed less than 6% of the flood route.

These spatial variations in local hydraulic conditions were profoundly important in controlling the distribution of flood processes and the resulting flood features. The deposition of tractively transported cobbles and boulders (measured clast diameters ranged from less than 10 cm to more than 10 m) were deposited in reaches of decreasing flow competence within quantitatively definable limits. The hydraulic conditions at areas of erosion were more difficult to precisely evaluate; however, erosion of basalt bedrock was primarily in reaches that had stream power magnitudes that exceeded 20,000 watts·m^{-2}. Cavitation was probably an important erosional process in areas of most intense flow conditions.

INTRODUCTION

The Bonneville Flood

The Bonneville Flood in southern Idaho provides an excellent opportunity to attempt to understand an important geomorphic event and associated geologic processes in both a qualitative and quantitative manner. A huge discharge associated with cataclysmic lowering of Pleistocene Lake Bonneville had profound impact on many sections of the Snake River drainage from Red Rock Pass to Lewiston, Idaho (Fig. 1). Abundant and well-preserved erosional and depositional features along the flood route have allowed for a well-founded qualitative understanding of the flood and flood processes (to a large extent, already docu-

O'Connor, J. E., 1993, Hydrology, Hydraulics, and Geomorphology of the Bonneville Flood: Boulder, Colorado, Geological Society of America Special Paper 274.

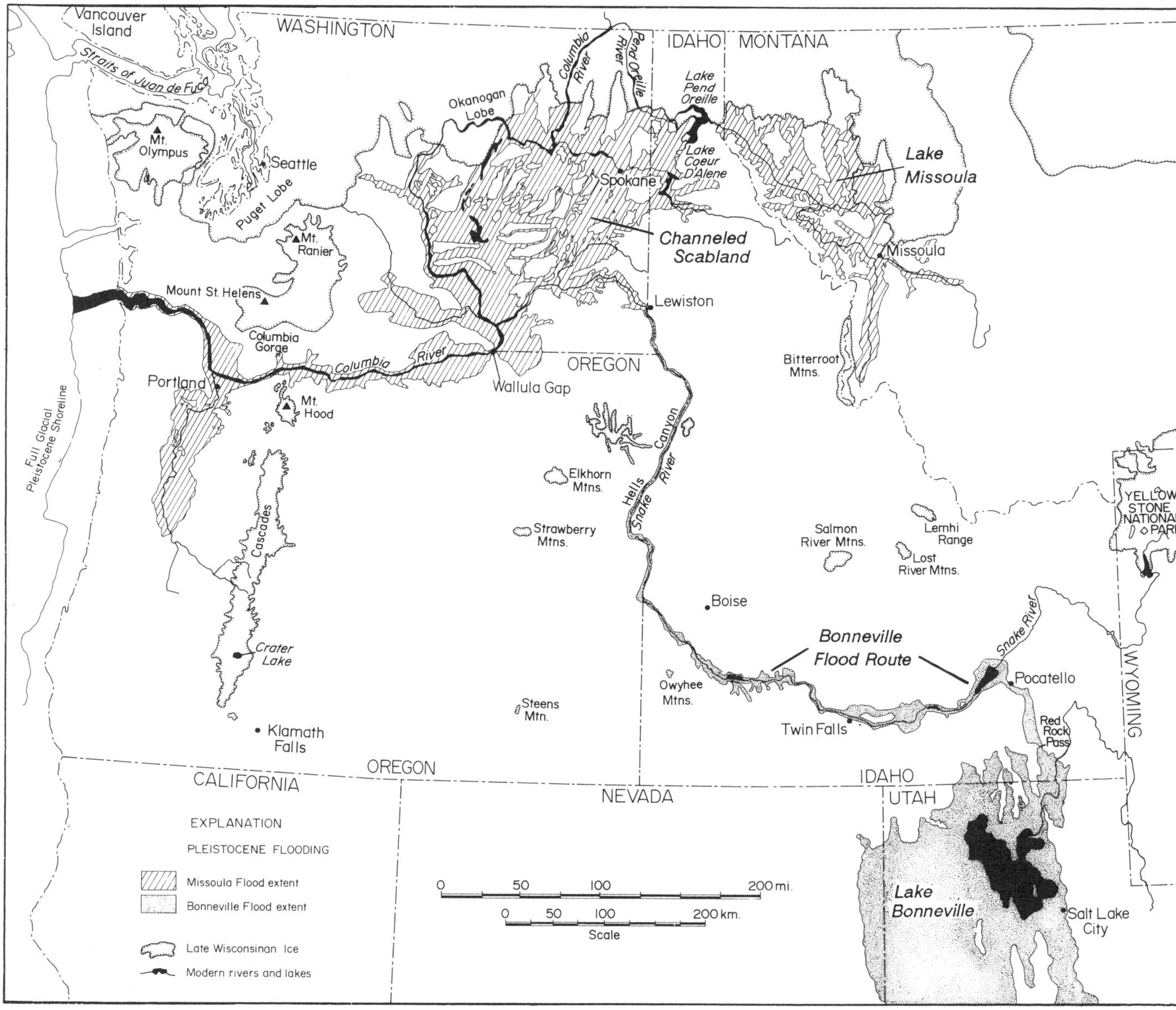

Figure 1. Regional map showing some aspects of the late Pleistocene paleogeography of the Pacific Northwest. In addition to Lake Bonneville and the route of the Bonneville Flood, the area inundated by Glacial Lake Missoula and the Missoula Flood(s) is shown. Modified from Baker (1983).

mented by previous studies). The existing framework of this qualitative understanding of the flood provided the impetus to analyze certain flood features and processes in terms of physical and engineering concepts and methods to achieve quantitative measures of some important flood processes.

Project scope

The objectives of the study were to: (1) use geologic evidence of maximum flood stages to define as precisely as possible the water-surface profile for the entire flood route between Red Rock Pass and Lewiston; (2) in suitable reaches, combine the geologic evidence of peak flow stages with engineering methods of flow modeling to determine peak discharges along the flow route and to obtain quantitative measures of local hydraulic parameters such as flow depth, velocity, shear stress, and stream power; and (3) compare these estimates of local hydraulic parameters with mapped and measured patterns of erosion and deposition to determine erosional and depositional thresholds.

Several aspects of the Bonneville Flood facilitated this study. Most importantly, there has apparently been only one flood and it was fairly simple hydraulically. The release of Lake Bonneville was, in relative terms, fairly slow, probably occurring over weeks or tens of weeks. The flood was confined to a single channel for most of its reach. Major flood-related channel change was restricted to isolated and identifiable locations. All of these factors contrast with the more studied, but more complex, history of multiple releases of Glacial Lake Missoula that carved the anastomosing channel network of the Channelled Scabland in eastern Washington (Fig. 1). Also important is that flood features are abundant and well preserved, and that high-level deposits and erosional features are numerous enough so that fairly tight constraints can be placed on the water-surface profile for many reaches. Low-level deposits and erosional features are also abundant and well exposed so that their relations to local flow hydraulics can be investigated.

This volume is organized in a manner following the project scope outlined above. First, cursory sections describe pertinent previous work in the area, the general geologic setting, and mapping of flood features and maximum stage evidence. The main body of the report consists of two sections describing (1) the use of flow modeling techniques in conjunction with the field evidence to obtain a quantitative understanding of the important hydraulic parameters for 10 reaches along the flood route; and (2) the relationship of these hydraulic parameters to mapped and measured erosional and depositional features.

PREVIOUS STUDIES

More than a century ago, G. K. Gilbert (1890) published his monograph on Pleistocene Lake Bonneville. Several pages of his report are devoted to describing conditions at the Lake Bonneville outlet near Red Rock Pass (Gilbert, 1890, p. 172–180), mostly the result of fieldwork performed between 1876 and 1879 (Gilbert, 1878; see also Hunt, 1982) (Fig. 2). In evaluating erosional features immediately downstream from the spillover point, Gilbert (1890, p. 176) documented evidence for the existence of a "Bonneville river" that resulted from the sudden release of Lake Bonneville and the consequent lowering of the lake. He suggested (p. 175) that "for a time the outpouring was a veritable debacle," perhaps with a discharge of "the flood volume of the Missouri" (p. 177). Gilbert, in his field notes (Hunt, 1982), described many of the effects of the flood between Red Rock Pass and Pocatello (Fig. 3) and clearly had an appreciation for the possible magnitude of the flood, suggesting that erosional features 400 ft above the Marsh Creek floodplain "accord" with the hypothesis of the Bonneville river.

The most spectacular and ubiquitous flood deposits are on the Snake River Plain in south-central Idaho (Fig. 3). Despite Gilbert's well-published observations, as well as two major U.S. Geological Survey (USGS) projects on the Snake River Plain in

U. S. GEOLOGICAL SURVEY LAKE BONNEVILLE PL. XXIX

RED ROCK PASS, THE OUTLET OF LAKE BONNEVILLE, AS SEEN FROM THE NORTH.

Figure 2. Lithograph from G. K. Gilbert's (1890) Lake Bonneville monograph illustrating a view south from the surface of the Marsh Creek Fan (Fig. 6). The Red Rock Pass constriction is near the bedrock outcroppings. The hummocky terrain of the Red Rock landslip is visible to the southwest across the flood-entrenched channel of Marsh Creek.

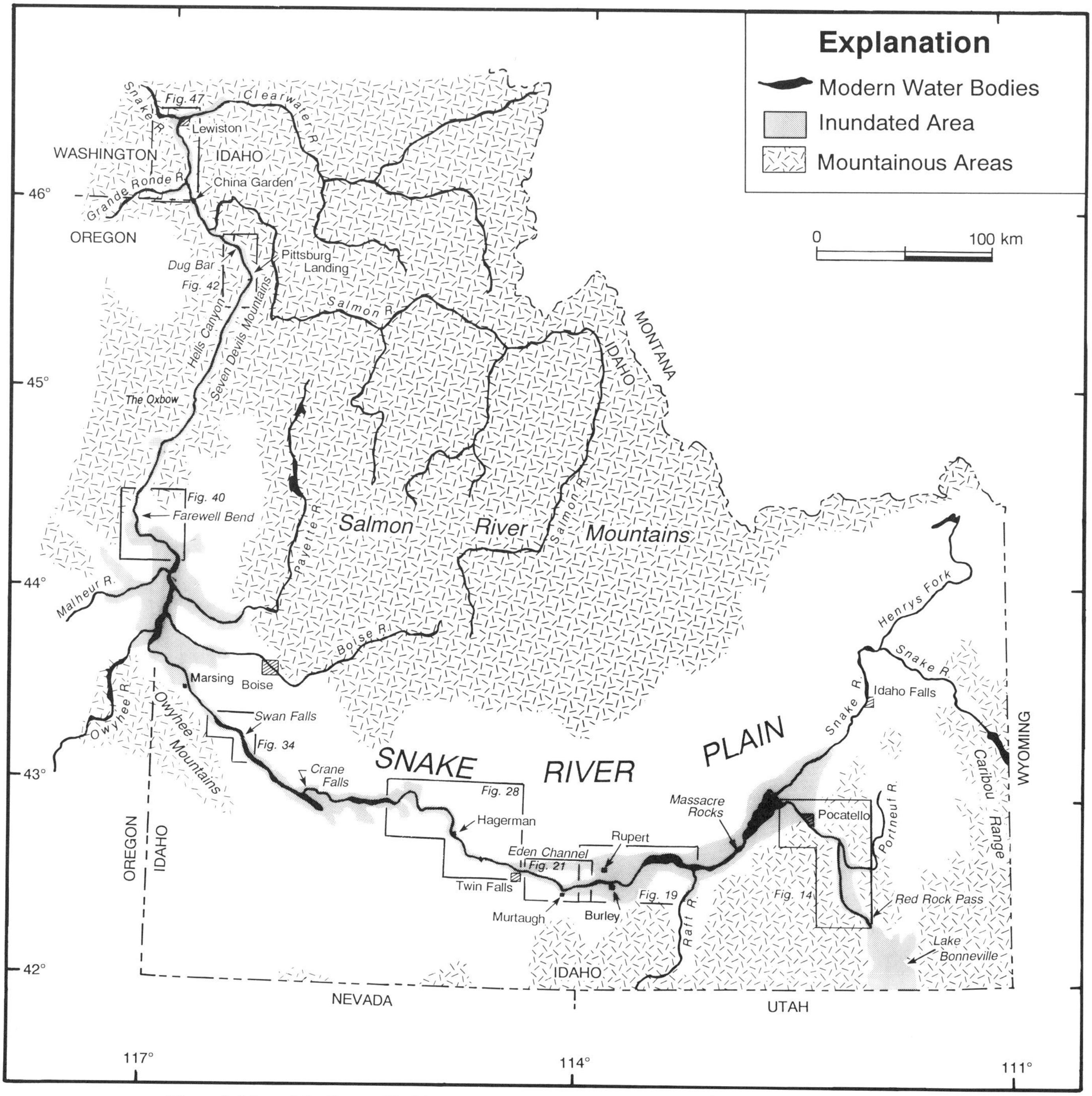

Figure 3. Map of the Bonneville Flood route and approximate area of inundation between Red Rock Pass and Lewiston. Areas covered by the study-reach maps (Figs. 14, 19, 21, 28, 34, 40, 42, 47) are also shown.

the early part of this century (Russell, 1902; Stearns and others, 1938), it was not recognized that the immense gravel accumulations and erosional tracts flanking the Snake River were related to the overflow of Lake Bonneville until Harold A. Powers and Harold E. Malde of the USGS began mapping Quaternary deposits on the Snake River Plain in the mid-1950s (Malde, 1968, 1987, personal communication).[1] By the early 1960s, several geologists working along the Snake River had recognized or reinterpreted features that could be attributed to the Bonneville Flood. Malde (1960) first published a short article outlining evidence for the flood along the Snake River, and then a full report in 1968 (Malde, 1968). He (with Powers) also produced a series of maps and reports on the Quaternary history of the western Snake River Plain that included Bonneville Flood features (Malde and Powers, 1962 [where Bonneville Flood deposits were assigned the formation name of "Melon Gravel"], 1972; Malde and others, 1963). D. E. Trimble and W. J. Carr, working in the Pocatello and American Falls areas, recognized the "Michaud Gravel" as a large delta or bar complex deposited by the Bonneville Flood; they also documented other flood-related features in the eastern Snake River Plain (Trimble and Carr, 1961a, b; Carr and Trimble, 1963). Stearns (1962) and Stearns and Anderson (1966) described Bonneville Flood features in the western Snake River Plain, and also, in conjunction with dam-site studies, evidence of the flood farther downstream in Hells Canyon.

The most complete of these analyses of Bonneville Flood features was a USGS Professional Paper by Malde (1968), which gives a detailed description of flood features for a large portion of the flood route on the Snake River Plain. Through careful mapping of the distribution of flood features, Malde was able to define the water-surface profile for about 300 km of the flood route between Marsing and the city of Twin Falls.[2] Furthermore, Malde linked, in a qualitative manner, patterns of erosion and deposition to local hydraulic and geologic conditions. That observation, and Malde's encouragement, provided the key impetus for the aim of this new look at Bonneville Flood features: to build on the qualitative understanding of the Bonneville Flood by quantitatively evaluating the hydraulic conditions associated with patterns and processes of erosion and deposition.

Since Malde's publication there has been little published work specifically about the Bonneville Flood; more recent information is mostly the byproduct of more general geologic investigations. Continued geologic mapping by Malde (1989a–e) has provided more information regarding the distribution of Bonneville Flood deposits in the western Snake River Plain and has led to a revised peak discharge estimate at Swan Falls (Jarrett and Malde, 1987) (Fig. 3). Webster and others (1982) identified Bonneville Flood deposits in Hells Canyon and downstream to Lewiston. Kehew (1977) also described deposits in the Lewiston area and built a strong case for their having been deposited by the Bonneville Flood. A surficial geologic map by Scott (1982) provides a compilation of surficial geologic mapping by previous workers, as well as new mapping that shows the distribution of Bonneville Flood sediment in the eastern Snake River Plain. Geologic mapping by Trimble (1976) delineated Bonneville Flood deposits in the Pocatello area. Bright and Ore (1987) summarized flood features and deposits between Red Rock Pass and Pocatello. Recent studies regarding the chronology of Lake Bonneville (see Currey [1990] for a recent review), as well as conditions at the outlet (Shroder and Lowndes, 1989), have important implications regarding the Bonneville Flood.

This project has benefited from geologists presently mapping along the Snake River who have shared their unpublished observations regarding Bonneville Flood features. Bill Bonnichsen and Margaret Jenks of the Idaho Geological Survey (IGS) have mapped flood deposits in the Swan Falls area. Kurt Othberg, also of the IGS, has mapped Bonneville Flood deposits in the lower Boise River valley. Tracy Vallier of the USGS has mapped extensively within Hells Canyon and has noted the locations of many Bonneville Flood deposits.

GENERAL SETTING AND CHRONOLOGY

The Bonneville Flood resulted from the filling and the consequent spillover of Pleistocene Lake Bonneville into the Snake River drainage about 14,500 yr B.P. About 4,750 km^3 of water was released near Red Rock Pass, Idaho, and traveled down the present courses of Marsh Creek and the lower Portneuf River before entering the Snake River Plain north of Pocatello. For the most part, the flood route followed the present westward course of the Snake River within the Snake River Plain through a series of canyons and basins before turning north and entering the narrow and deep Hells Canyon along the Idaho-Oregon boundary. At Lewiston, the Snake River turns abruptly to the west and enters the Columbia River near Pasco, Washington. The scope of this study was restricted to upstream of Lewiston (Fig. 3).

Lake Bonneville

During the latest Pleistocene, Lake Bonneville was the largest of numerous lakes within the closed basins of the Basin and Range province of western North America (Benson and Thompson, 1987), at its highest level covering 51,530 km^2 in western Utah and parts of Nevada and Idaho (Fig. 4). As reported by Currey (1990), during its last cycle, Lake Bonneville began to rise episodically from about 28 ka until it reached its maximum,

[1]This is especially curious, because I. C. Russell, who published a monograph on the geology and water resources of the Snake River Plain, was a major participant in G. K. Gilbert's studies of Lake Bonneville.

[2]The city of Twin Falls is distinguished from physiographic Twin Falls (the actual falls).

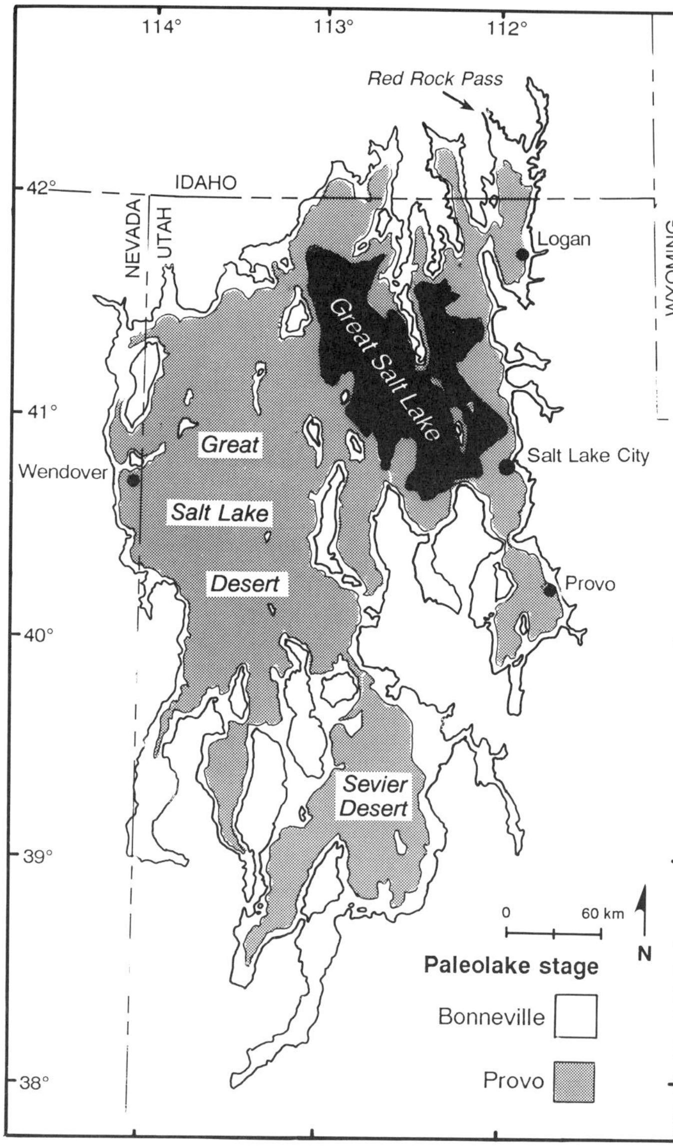

Figure 4. Map of Bonneville and Provo levels of Lake Bonneville and the present area of Great Salt Lake. Overflow of the divide at Red Rock Pass and the resulting Bonneville Flood was responsible for the drop from the maximum Bonneville level to the Provo level. Modified from Currey (1990).

Bonneville, level at about 15.2 ka (Fig. 5).[3] At levels below the Bonneville shoreline complex [1,552 m (5,092 ft)[4] near Red Rock Pass], the lake had no outlet and the water level was controlled by the balance between precipitation and evaporation. When the lake surface reached an altitude of 1,552 m (5,092 ft), it met the lowest of the divides between the Bonneville basin and the watershed of the Snake River, near Red Rock Pass, Idaho, and began to drain to the north (Gilbert, 1878, 1890). Lake Bonneville was apparently maintained at an altitude near 1,552 m (5,092 ft) for several hundred years until about 14.5 ka (Scott and others, 1983; Currey and others, 1983a; Currey and Oviatt, 1985; Currey, 1990) (Fig. 5), time enough for the prominent Bonneville shoreline complex to develop and for the short regression between 15.0 and 14.5 ka (the Keg Mountain oscillation of Currey and others [1983b]).

At about 14.5 ka, conditions at the outlet changed and rapid incision began, resulting in a 108-m drop of the lake level before the outlet stabilized at an altitude of 1,444 m (4,738 ft). The lake remained at this level (the Provo level; Figs. 4, 6), with continued overflow into the Snake River drainage, until about 14.2 ka when the lake rapidly regressed to Holocene levels similar to that of present-day Great Salt Lake (Scott and others, 1983; Currey and others, 1983a; Currey, 1990).

Red Rock Pass

The Bonneville Flood was the result of rapid incision of the preflood divide between the Bonneville basin and the Snake River drainage at a point just north of Red Rock Pass (Gilbert, 1878, 1890; Shroder and Lowndes, 1989; Currey, 1990) (Figs. 2, 6). The lake threshold prior to the flood was a large alluvial fan, which emanated from the Marsh Creek drainage and the Portneuf Range to the east and impinged against alluvial fan deposits

[3]All chronologic information presented here is that reported and compiled by Currey and his associates (see Currey [1990] for a review) and is based on radiocarbon dates on materials that can be stratigraphically linked to shoreline features in the Bonneville basin.

[4]A note on units: because all original mapping and flow modeling was performed with English-unit maps (USGS 7.5′ quadrangles), all altitudes are reported in both metric and English units; all other measurements are reported only in metric units.

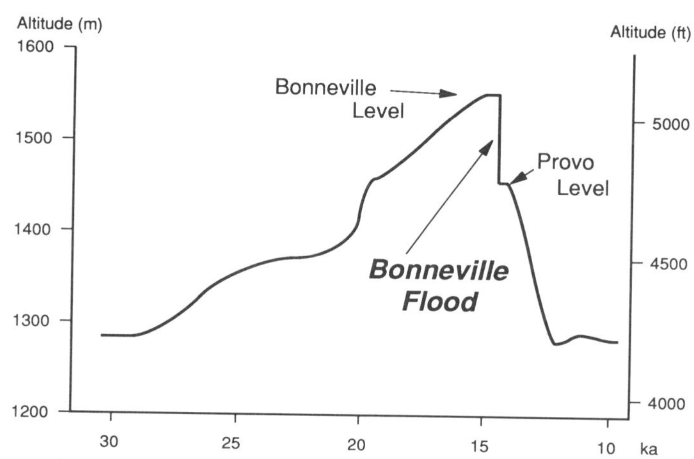

Figure 5. Generalized late Pleistocene history of lake levels in the Lake Bonneville basin between 30 and 10 ka. The only time of open-basin conditions was while the lake was at the Bonneville and Provo levels. Subsequent to stabilization at the Provo level after the Bonneville Flood, changing climatic conditions resulted in rapid desiccation to levels similar to present Great Salt Lake. After Currey (1990).

or bedrock of the Bannock Range to the west. Faunal evidence indicates that the Marsh Creek fan was active during the late Quaternary (Bright and Ore, 1987). The fan is dominantly composed of coarse, well-rounded gravel and cobbles, apparently fluvially deposited.

Once surface overflow of this natural dam began, it probably eroded quickly because of the unconsolidated nature of the fan deposits, the large hydraulic head of the impounded lake, and the steep gradient of the fan surface to the north. Downcutting of the divide continued until the flow met bedrock at the constriction near Red Rock Pass (Gilbert, 1890, p. 175). The bedrock divide is about 9 m (30 ft) below the present surface at Red Rock Junction and is near the outlet altitude for the Provo lake level (Williams and Milligan, 1968). Contemporaneous with the erosion of the divide and the release of flood water were several landslides (Bright and Ore, 1987; Shroder and Lowndes, 1989). The largest of these was the Red Rock landslip that involved about 17 km^2 of the Bannock Range on the west side of Marsh Valley northwest of Red Rock Pass.

Ground-water seepage through the fan and the consequent development of sapping channels on the fan surface may have led to initial failure of the alluvial barrier at Red Rock Pass. There was more than 75 m of hydraulic head between the maximum Lake Bonneville altitude and the fan surface at the northern, downstream, end of the fan. This hydraulic gradient probably led to substantial ground-water flow through the fan. Moreover, at points where ground water intercepted the fan surface, small sapping channel networks may have formed. As these channels eroded back up the fan surface, one may eventually have connected with the lake and allowed rapid surface overflow and erosion to begin. An extensive network of meandering channels, presently inactive, is preserved on the unflooded portion of the Marsh Creek fan. These networks all head at or below 1,555 m (5,100 ft) in small, scallop-shaped depressions in a manner that strongly resembles sapping channel networks (Dunne, 1980; Laity and Malin, 1985), suggesting that this process was operating during times of maximum lake levels. Furthermore, this type of failure mechanism is more consistent with the complicated and prolonged lake-level history at maximum Lake Bonneville levels than is the model of failure by simple overtopping. A prolonged period of substantial subsurface drainage helps explain the several-hundred-year hiatus between Lake Bonneville reaching its maximum stage and the subsequent failure.

Flood route

Downstream of Red Rock Pass, the flood route followed the valleys of Marsh Creek and the lower Portneuf River northward until entering the Snake River Plain north of Pocatello. For most of this reach, the inundated area was several kilometers wide and bounded by alluvial fan deposits flanking the Portneuf and Bannock Ranges (Fig. 7). An important exception is at the Portneuf Narrows, where the Portneuf River cuts through the Bannock Range about 10 km southeast of Pocatello. This bedrock constriction apparently influenced the flood profile for several tens of kilometers upstream.

The flood entered the Snake River Plain north of Pocatello and followed this vast volcanic plain westward for about 600 km until turning north and entering Hells Canyon (Fig. 3). Malde (1991) has provided a recent review of the Quaternary geologic history of the Snake River Plain, which offers the following description of the general topography (p. 251):

> The Snake River Plain is an area of subdued, locally almost featureless, relief surrounded by mountains and highlands. The plain is as much as 70 km wide in the western part, 90-100 km wide in the eastern part, and about 600 km long. Its surface rises with a gradually increasing gradient from altitudes of less than 700 m at the western end to about 2,000 m at the eastern end.

The Snake River is the primary topographic feature on the plain, and its canyons and valleys were the major conduit for the flood.

For most of the Snake River Plain, the Snake River is deeply incised, as much as 200 m below the level of the basaltic upland. This incision is partly the result of base-level fall caused by integration of the Snake River drainage into the Columbia River system. Prior to integration, the western Snake River Plain was a depositional center characterized by low-energy fluvial and lacustrine environments (Malde, 1991). Wheeler and Cook (1954) stated that the Snake River was probably captured near the head of Hells Canyon near "the oxbow" (Fig. 3). Available biostratigraphic and radiometric age data indicate that capture occurred between about 1.8 and 2.0 Ma and resulted in a more vigorous, through-going drainage (summarized by Malde [1990]).

Both the present course and character of the Snake River within the Snake River Plain are the result of this base-level fall and complex interactions between extensive Quaternary volcanism and the river. Voluminous basaltic volcanism has displaced the Snake River generally south toward the southern margin of the plain (Malde, 1971, 1991; Covington and others, 1985). At places where the river was diverted to locations of erodible sediment, wide valleys with low gradients have formed, generally bounded on the north by basaltic rims. For reaches where the river has been forced to cut through volcanic rocks, high-gradient narrow canyons have formed (Fig. 8). Locally, the Snake River was dammed by basalt flows and formed temporary lakes (Stearns and others, 1938, p. 89; Malde, 1971, 1982, 1991; Carr and Trimble, 1963; Scott and others, 1983).

As a result of this diverse geologic and geomorphic setting, the Bonneville Flood coursed through different hydraulic environments as it flowed across the Snake River Plain. Reaches of ponded or slow-moving flow were separated by reaches of intensely energetic conditions. Despite this complexity, the flood route along the Snake River Plain has been the most amenable for detailed study because of the general diversity and abundance of flood features, as well as because of their excellent exposure and preservation.

At the western edge of the Snake River Plain, near Farewell

U. S. GEOLOGICAL SURVEY LAKE BONNEVILLE, PL. XXVIII

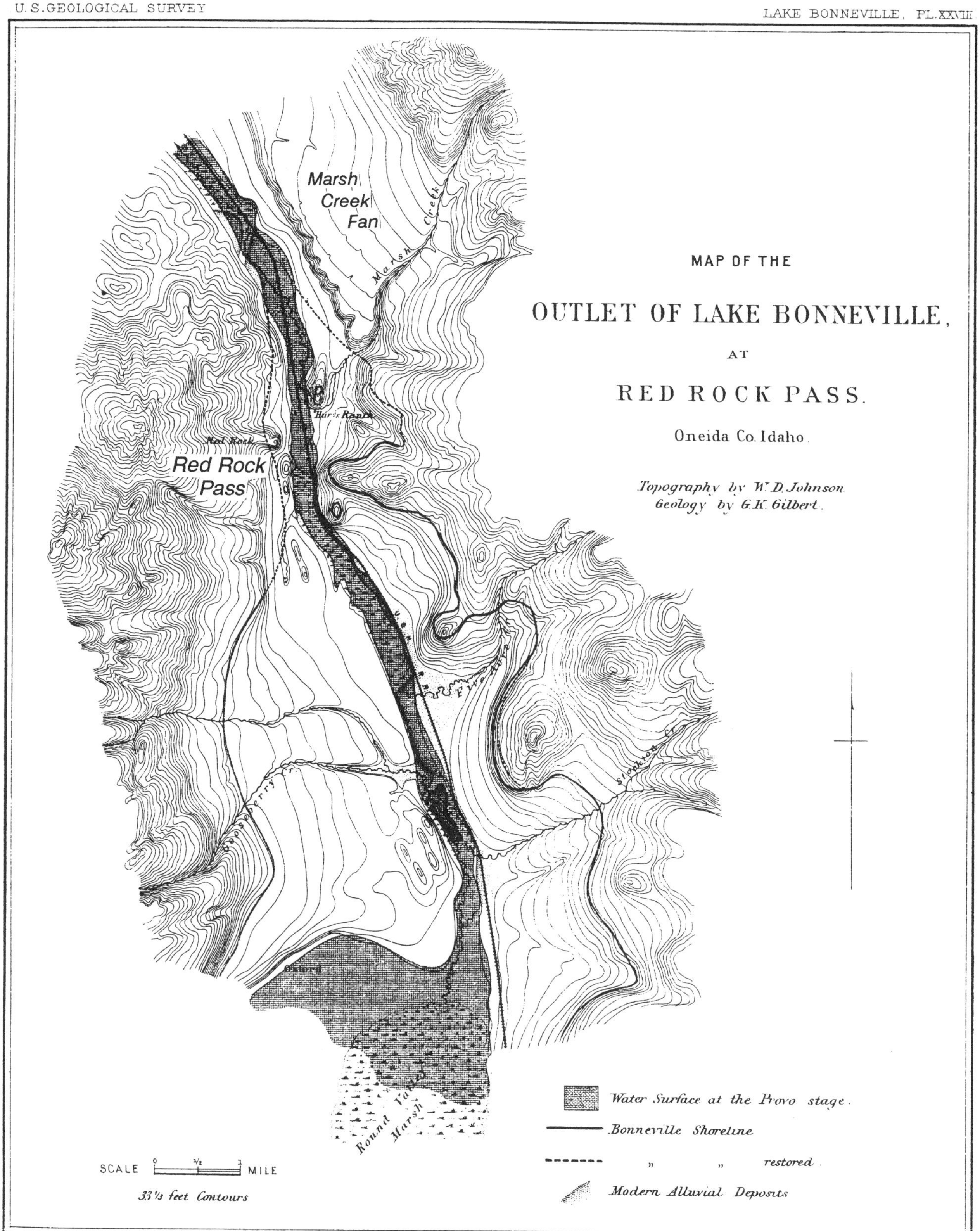

Julius Bien & Co. lith. Drawn by G. Thompson

Figure 7. Oblique aerial view to the east of a portion of the Portneuf Valley immediately upstream of the Portneuf narrows (Figs. 13, 14). The town of Inkom is in the upper part of the view. Portions of the basalt flows on the valley bottom were severely eroded by the flood, leaving now-dry cataracts and scoured and fluted surfaces (Fig. 17). The prominent scarps truncating the piedmont gravel deposits, forming the north valley margin, were probably cut during the flood. There is no evidence for water of the Bonneville Flood having flowed across the piedmont surfaces. Consequently, the altitudes of the tops of the scarps are inferred to be maximum limits on the maximum flow stage. Flow was toward the camera.

Figure 6. Topographic map of Red Rock Pass area by G. K. Gilbert (1890) illustrating relationships near the Lake Bonneville outlet. Prior to the flood, the divide between the Lake Bonneville basin and the Snake River watershed was formed by the Marsh Creek Fan where it impinged against the mountains to the west. The present divide is about 2 km south at Red Rock Pass. Note that stream labeled "Five Acre Creek" is now known as Swan Lake Creek and most of the area in the map is now in Bannock County, Idaho.

Bend, the Snake River turns northward and enters the deep, narrow gorges that characterize the river as it follows the Idaho-Oregon and Idaho-Washington borders to Lewiston (Fig. 3). This 300-km section of river is remote and rugged, culminating in Hells Canyon, one of the deepest gorges in North America. Valley bottoms are narrow, generally less than 1 km wide, and dissected mountainous terrain rises steeply from both sides to altitudes of several thousand meters. The river is steep, an attribute exploited for power generation and river running.

Within this reach, the Snake River has cut into two eastern arms of basalt flows of the Miocene Columbia River Basalt Group: the basalt flows of the Weiser embayment that form the uplands surrounding the river from Farewell Bend to near Oxbow Dam (Fitzgerald, 1982), and the basalts of the Clearwater embayment that are exposed in the northern part of the reach from near the Salmon River confluence to Lewiston (Reidel, 1982). Between these two basaltic embayments is Hells Canyon proper, where the Snake River is especially narrow and steep as it cuts through the Seven Devils and Wallowa Mountains (Fig. 9). These ranges consist of Permian-Triassic volcanic and volcaniclastic units of the Seven Devils Group and associated ophiolitic sequences. These units are overlain by Jurassic clastic rocks and have been intruded by Mesozoic granitic rocks (Vallier, 1977).

The high gradient and relief along the Snake River between Farewell Bend and Lewiston result from the resistant rock types and the recent integration of the river into the Columbia River drainage. If the events outlined above regarding the capture of the Snake River are correct, then a major portion of the incision of Hells Canyon has probably been accomplished during the last 2 million years.

The narrow valleys and active hillslope processes have resulted in few preserved and exposed flood deposits within Hells Canyon. There are only isolated locations where the canyon was wide enough to allow for substantial deposition along the valley bottom. Small and high-level marginal deposits are more abundant, preserved in tributary mouths and canyon reentrants that have escaped subsequent erosion. At the downstream end of the study area, near Lewiston (Fig. 3), the valley widens and channel gradient diminishes. Consequently, the flood deposited large bars consisting of material eroded from Hells Canyon. Downstream of Lewiston, Bonneville Flood features are not readily distinguishable from, or have been buried by, deposits of subsequent Missoula Floods.

GEOLOGIC EVIDENCE OF THE BONNEVILLE FLOOD

Physical features left by the Bonneville Flood were examined for three purposes: (1) to define as precisely as possible the maximum stages achieved by the flood for as much of the flood route as possible; (2) to map patterns of erosion and deposition associated with different hydraulic environments; and (3) in conjunction with the first two goals, to determine reaches of the flood route that are suitable for flow modeling and quantitative analysis of depositional and erosional processes.

Figure 8. View west (downstream) near the Rock Creek constriction (photograph location shown in Fig. 31). The basalt-walled canyon of the Snake River is approximately 120 m deep at the constriction, which is visible in the middle distance. The gravel bar in the foreground (with the tire tracks on it) is inferred to have been deposited by the Bonneville Flood, indicating that the canyon was brimful at peak discharge.

Figure 9. View south (downstream) from near Dug Bar within Hells Canyon (photograph location shown in Fig. 44). The Snake River completely occupies the canyon bottom, and the canyon margins are steep and irregular. Photograph taken from a position near maximum flood stage about 115 m above the present river level.

Maximum flood stages

Several lines of evidence were used to constrain the flood profile as precisely as possible. Flood deposits and erosional features provide minimum estimates of maximum flood stages. Although more difficult to interpret with confidence, altitudes of locations that are inferred not to have been flooded serve as maximum constraints on flood stages. The types of features used for defining the water surface of the Bonneville Flood include maximum altitudes of eddy and slackwater deposits, and the upper limits of scabland topography and basalt stripped of its loess cover. Tops of erosional scarps in loess and older alluvium were used as maximum constraints on the flood stage.

The altitudes of flood features used to constrain maximum flow stage were generally determined by surveying from benchmarks, checked altitudes (labeled spot altitudes on prominent points, intersections, and section corners on USGS topographic maps), or reservoir levels of known water-surface altitude. At locations where contour intervals were small or where it was impractical to survey, altitudes were determined as precisely as possible by mapping onto topographic maps. The estimated accuracy of the determined altitudes depends on the method of survey or the contour interval, and in the cases of surveys from unmonumented checked altitudes (such as road intersections where altitudes are given on USGS maps), the ability to relocate the surveyed point.

General comments regarding the results of mapping

Figure 10 provides an overview of the flood profile within the project area. In a general manner, the water-surface profile of the Bonneville Flood follows the gradient of the Snake River (and the Portneuf River and Marsh Valley). In detail, however, much of the drop in the profile was in short reaches of high gradient that were separated by longer reaches of relatively low water-surface gradient. These steps in the water-surface profile were caused by flow constrictions that hydraulically impounded flow for several tens of kilometers upstream. Some of the important constrictions in controlling the flood profile were at the Portneuf Narrows, Massacre Rocks, Milner Dam, Rock Creek, Swan Falls, and Farewell Bend. In reaches of generally steeper gradient (within Hells Canyon and near Murtaugh), the flood profile was probably also stepped. However, the backwater from constrictions in these reaches was not as extensive; consequently, the lengths of impounded reaches were much shorter, and the stepped nature of the water-surface profile is not so well defined (Fig. 10).

Most of the mappable deposits and erosional features are

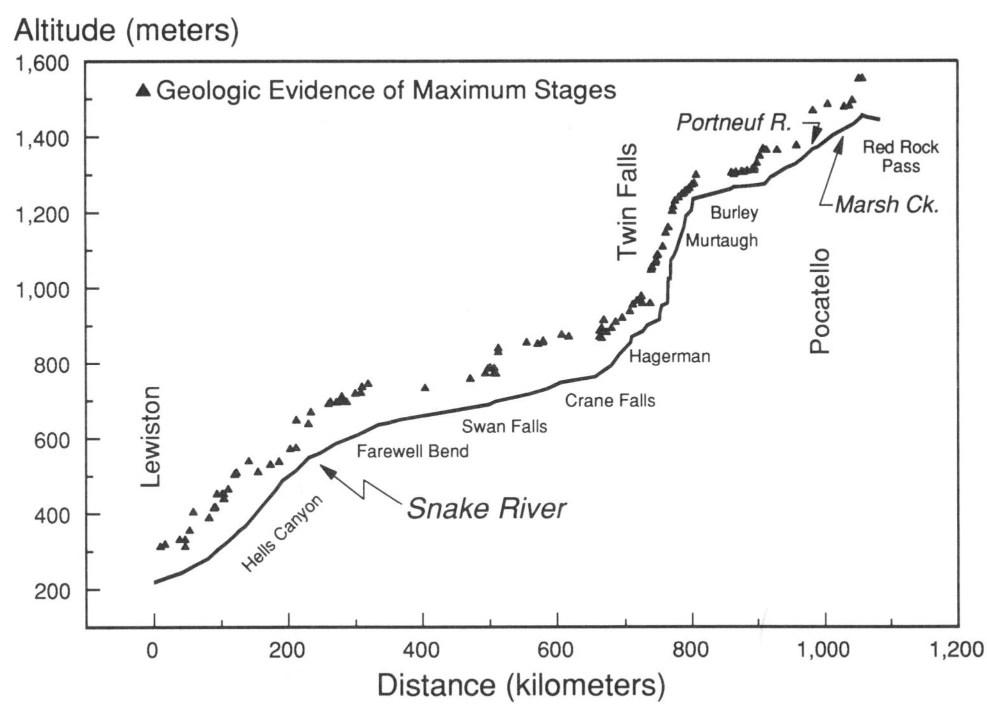

Figure 10. Profiles of the Snake River, Portneuf River, and Marsh Creek with evidence of maximum Bonneville Flood stages. A summary of the evidence for maximum flood stages is given in Appendix 2. The altitudes plotted here are minimum estimates of measured altitudes for flood features. As discussed in the text, the Bonneville Flood profile was stepped with much of the water-surface drop over short distances separated by long reaches of more quiescent flow. Evidence for maximum stages in the Eden channel (north of the Snake River) is not shown.

concentrated near these constrictions and reaches of high gradient where there were intense flow velocities and energy expenditures. The reaches of low water-surface gradient are generally devoid of flood features with the exception of fine-grain sediment that was deposited in backwater areas.

Along some reaches, the Bonneville Flood tremendously changed the landscape. In the vicinity of Massacre Rocks (Fig. 3), for example, a lava dam blocking the preflood course of the Snake River was breached, resulting in the emptying of Pleistocene American Falls Lake and entrenchment of a new course of the Snake River (Trimble and Carr, 1961b; Scott and others, 1982). Near the city of Twin Falls, the Snake River canyon was substantially enlarged by flood flow from the adjacent uplands spilling over the canyon rims and eroding large and spectacular cataract complexes (Fig. 11).

HYDRAULIC ANALYSIS

Step-backwater hydraulic modeling in conjunction with the geologic evidence of maximum Bonneville Flood stages provided a means to evaluate flow conditions quantitatively for several reaches of the flood route. In addition to providing estimates of peak discharge for each of the modeled reaches, the step-backwater calculations allow for estimation of local hydraulic conditions and their spatial variation. Key flow properties that can be locally determined include variables such as energy slope, velocity, shear stress, and stream power.

Step-backwater modeling

Step-backwater calculations are premised on the principles of conservation of mass and the conservation of energy associated with steady, gradually varied flow. For a series of representative cross sections along a channel, an energy-balanced water-surface profile can be calculated that is a function of discharge, flow resistance, and channel geometry. For paleohydrologic applications, the calculated profiles are compared to the geologic evidence of maximum stages in order to estimate the peak discharge of the flow that left the deposits. The precision of the resulting discharge estimate can be estimated by evaluating the quality of the match between the modeled profile and the profile defined by the geologic evidence. For the discharge and associated water-surface profile that provides the best fit, the local hydraulic conditions can be estimated from the solution at each of the modeled cross sections. For this study, a computer program (HEC-2) developed by the Hydrologic Engineering Center (1985) of the U.S. Army Corps of Engineers was used for the calculation of the water-surface profiles.

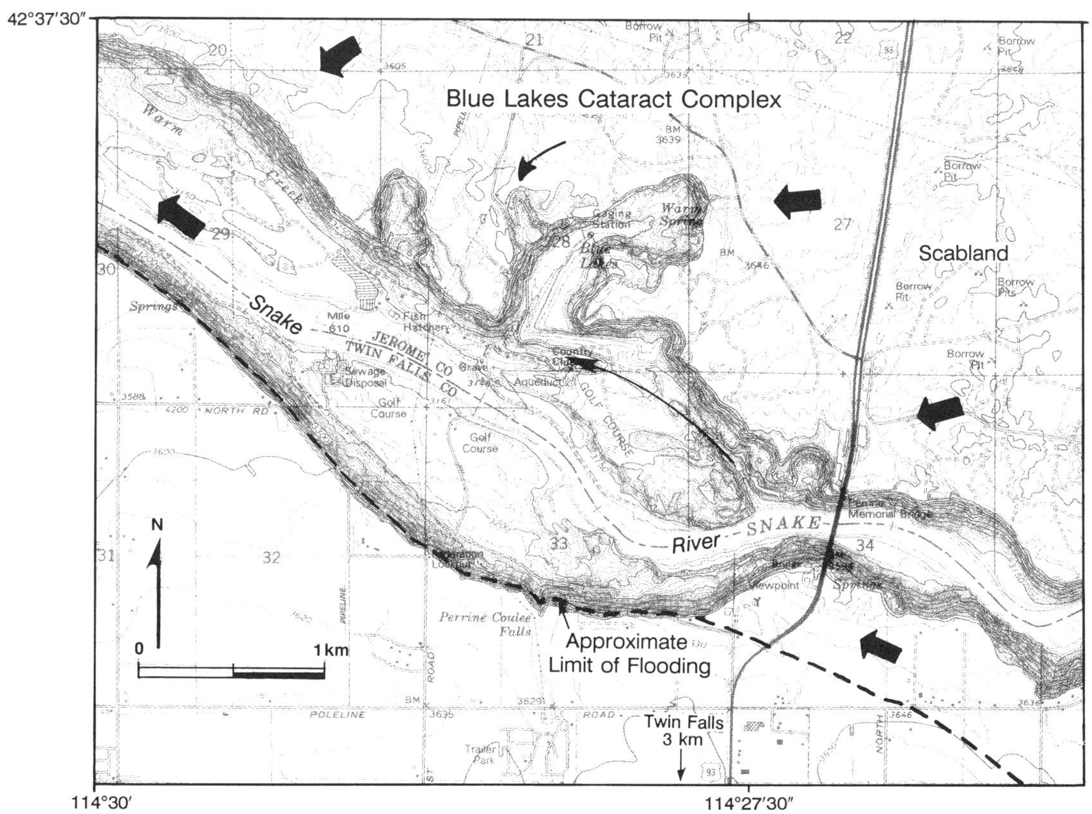

Figure 11. The Blue Lakes cataract complex north of the city of Twin Falls. These cataracts were formed by flood water in the Eden channel spilling back into the canyon of the Snake River from the north. The arrows indicate general flow directions interpreted from the orientations of flood features. Topography from the Twin Falls (ID) 7.5-min USGS Quadrangle.

Step-backwater calculations of water-surface profiles have been used extensively by the engineering community for floodplain hazard analysis for several decades (Chow, 1959; Feldman, 1981), but it has been only recently recognized this type of analysis can aid geomorphic and paleohydrologic studies (e.g., Ely and Baker, 1985; Webb, 1985; O'Connor and others, 1986; Jarrett and Malde, 1987; Florsheim and Keller, 1989; Wohl 1992a, b).

The step-backwater method for modeling flow conditions of the Bonneville Flood was chosen for a variety of reasons. Foremost, it is a simple technique that describes the pertinent aspects of the flow conditions relating to the resultant geomorphology. Additionally, the assumptions intrinsic to the step-backwater method are straightforward and relatively easy to evaluate. More sophisticated hydraulic models exist, such as unsteady flow models, coupled flow and sediment transport models, and two-dimensional flow models. It is not clear, however, that the additional assumptions and complexity required in applying these more sophisticated treatments warrant their use for obtaining a first-order understanding of the hydraulic conditions of the Bonneville Flood. As will be shown, application of the step-backwater method seems to provide substantial insight into flood processes. Undoubtedly, more sophisticated treatments of the Bonneville Flood are possible and desirable, and would yield additional insights. It seems to me, however, that more complicated analyses would be most efficiently performed by building on results of relatively simple analyses and focusing on understanding aspects of the flow conditions that cannot be understood by simpler means.

Theory. Step-backwater calculations are premised on the principle of conservation of energy; the total mechanical energy at one cross section equals the mechanical energy at an adjacent cross section less any energy loss between the sections. Necessary assumptions for applying this principle with the step-backwater method for water-surface profile calculation are that flow conditions are (were) steady with time and gradually varied in space. Furthermore, to predict water-surface profiles associated with gradually varied flows, a necessary assumption is that "the head [energy] loss at a section [of gradually varied flow] is the same for a uniform flow having the velocity and hydraulic radius of the section" (Chow, 1959, p. 217). This allows the use of uniform-flow formulae to evaluate the energy slope at each cross section; that is, the uniform-flow friction coefficients (Mannings *n*) are assumed to be applicable for gradually varied flow conditions (Feldman, 1981). For these conditions, the one-dimensional energy equation is appropriate for solution of the flow profiles (Chow, 1959, p. 267) (see Fig. 12 for defining illustration);

$$z_1 + d_1 + \frac{\alpha_1 v_1^2}{2g} = z_2 + d_2 + \frac{\alpha_2 v_2^2}{2g} + h_e. \quad (1)$$

Equation 1 expresses conservation of mechanical energy for gradually varied flow within an incremental reach of channel. The sum of a flow's potential and kinetic energy must equal that of a downstream cross section less any energy (head) losses (h_e) between sections. Energy loss (mechanical energy converted to heat or work done on the flow boundaries) between sections is calculated as the sum of frictional losses (relating to channel roughness) and form losses (relating to channel expansions and contractions and associated velocity changes):

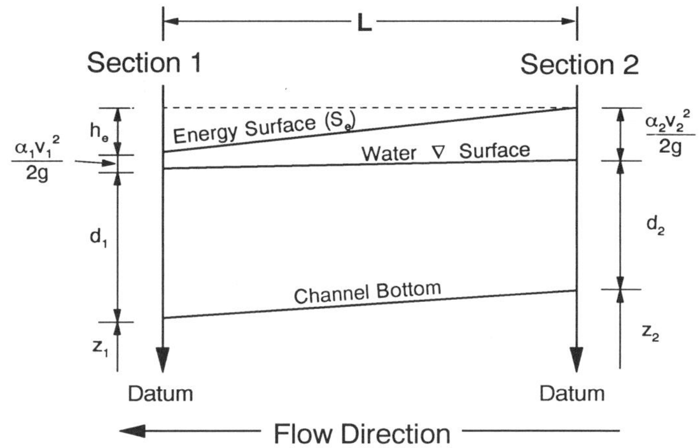

Figure 12. Definition diagram of conservation of mechanical energy for gradually varied flow (in this case, a contracting reach) for small channel slopes: z is the altitude of the channel above an arbitrary datum, L is the distance between the two cross sections, d is the flow depth, g is the gravitational acceleration, v is the flow velocity, α is the velocity-head coefficient accounting for nonuniform velocity distribution in a subdivided channel, and h_e is the head (energy) loss between cross sections. Note that, in natural stream systems, the water-surface slope may deviate locally from both the channel slope and the energy slope (S_e), depending on the local channel geometry and rates of energy loss.

$$h_e = L\bar{S}_f + C\left(\frac{\alpha_2 v_2^2}{2g} - \frac{\alpha_1 v_1^2}{2g}\right), \quad (2)$$

where C is the energy-loss coefficient relating to channel expansions and contractions, L is the distance between cross sections. The local friction slope (S_f) is calculated as a function of Mannings n,

$$S_f = \frac{n^2 v^2}{R^{4/3}}, \quad (3)$$

where R is the hydraulic radius of the channel. For known conditions (stage and discharge) at one cross section, equations 1 and 2 are simultaneously solved to calculate the water-surface altitude at the adjacent cross section. The newly predicted water-surface altitude is then taken as known for the next incremental step along the reach. In this manner a water-surface profile can be calculated for an indefinite length of channel provided stage and discharge conditions are given at an initial cross section and that the modeling assumptions are satisfied. Chow (1959) and Feldman (1981) have provided more detailed explanations of the solution of step-backwater profiles. The user's manual accompanying the HEC-2 program (Hydrologic Engineering Center, 1985) provides specific details regarding the computation procedure for the calculations performed in this study.

Application to natural channels. The assumptions intrinsic to step-backwater modeling impose constraints on their use in natural channels for obtaining meaningful results. The basic assumptions of steady, gradually varied flow require that for short channel increments:

> (1) The flow is steady, that is, the hydraulic characteristics of the flow remain constant for the time interval under consideration; and (2) that streamlines are parallel; that is, hydrostatic distribution of pressure prevails over the channel section. [Chow, 1959, p. 217]

In application, these assumptions require that: (1) the flow was steady enough so that the peak discharge was of sufficient duration to have simultaneously affected the entire modeled reach; and (2) channel cross sections are separated into short enough increments so that the flow characteristics do not vary significantly between cross sections. For complex channels and for other locations of rapidly varying flow conditions, one-dimensional gradually varied flow may not be a tenable assumption regardless of the lengths of the modeled channel increments, and, consequently, the calculated energy losses may not be accurate.

Modeling can be performed in either the subcritical or supercritical flow regimes; in practice the difference being that supercritical calculations must proceed in the downstream direction and subcritical calculations proceed in the upstream direction in order to maintain numeric stability (Chow, 1959, p. 265). Experience with numerous bedrock-confined rivers in the southwestern US (Ely and Baker, 1985; O'Connor and others, 1986; Partridge and Baker, 1987; Webb and others, 1988), as well as with the largest Missoula Flood(s) (O'Connor and Baker, 1992), has indicated that subcritical flow is predominant in many resistant boundary fluvial systems. Long reaches of subcritical flow may be separated by short reaches of critical or supercritical flow in channel constrictions (O'Connor and Webb, 1988). Modeling results in the neighborhood of transitions from supercritical to subcritical flow regime are unreliable because of the large energy losses associated with the hydraulic jumps that can occur in such locations.

A critical element for accurate step-backwater flow reconstruction is accurate description of the channel geometry. Cross sections must be spaced and placed so as to define all expansions, constrictions, and changes of slope so that the energy losses resulting from spatial variations in cross-section geometry and roughness are adequately determined. Regions of effective flow (flow in the downstream direction) must be distinguished from portions of the channel that do not convey downstream-directed flow. Regions of ineffective flow include areas behind bedrock spurs into the channel and in tributary mouths where there is flow separation and development of large-scale eddies. Because step-backwater calculations are particularly sensitive to channel geometry, reliable results can only be obtained when the channel geometry at maximum stage is known or can be approximated. For this reason, stable, bedrock-confined channels are most appropriate for this type of analysis.

Although discharge, initial flow depth, and energy-loss coefficients all must be specified to calculate a water-surface profile, discharge is the primary profile-controlling variable for a given set of cross sections. This is especially the case in confined channels where flow depths are relatively large with respect to flow width. Furthermore, flow depths are independent of energy-loss coefficients at sections of critical flow. Sensitivity analyses conducted for modeled Bonneville Flood profiles (Jarrett and Malde, 1987) and for bedrock fluvial systems of smaller and larger scales (O'Connor and Webb, 1988; O'Connor and Baker, 1992) indicate that discharge is indeed the most important variable controlling the modeled profiles. This is especially the case in reaches immediately upstream of sections of critical flow (Jarrett, 1984).

Application to the Bonneville Flood. Several reaches of the Bonneville Flood are particularly suitable for step-backwater flow reconstruction. Foremost, the hydrology and hydraulics of the flood were relatively simple. For most of its length, the flood route was confined to a single channel. Several reaches are confined by rock types that were resistant to flood erosion. Also, there is no evidence for major channel change after the flood. Consequently, for these reaches, the present topography is probably similar to that of maximum flood stages, and, in most places, cross sections measured from the present topography adequately represent the channel during peak stages. The flow was probably fairly steady. After the Marsh Creek Fan at Red Rock Pass was completely breached (probably early in the flow), maximum discharges would have been attained and maintained as the large volume of water slowly funneled through the Red Rock Pass constriction. Moreover, for many reaches, there is abundant geologic evidence of maximum stages, so that step-backwater modeling results can be compared to the geologic evidence of maximum flow stages to the degree necessary to produce accurate discharge estimates.

It is easier to define confidently minimum altitudes of maximum flood stages than it is to provide upper limits for maximum stages. Consequently, the strategy in determining the peak discharge for each of the study reaches was to provide a firm *minimum* estimate of the peak discharge. Toward this goal, the high-water indicators were usually directly compared to the computed water-surface profile. However, in energetic environments, where it is possible deposits were locally emplaced at higher altitudes than the average water-surface altitude (by run-up), the altitudes of the deposits were evaluated with reference to the calculated energy surface. Despite this approach of defining a minimum estimate of the peak discharge, estimates calculated here are thought to approximate closely the actual maximum discharge.

Flow conditions were modeled for 10 reaches between Red Rock Pass and Lewiston (Table 1). The major criterion in the selection of the reaches was their satisfaction of the assumptions intrinsic to step-backwater calculations. Additional important criteria included (1) the abundance and quality of the geologic evidence of maximum flood stages, (2) the abundance of geomorphic features that could potentially be related to the modeled

TABLE 1. MODELED REACHES AND SUMMARY OF RESULTS

Reach Name	Geographic Location	Snake River Mileage*	Number of Cross Sections	Reach Length (km)	Bracketing Discharges (10^6 m^3 · sec^{-1})	Utilized Estimate† (10^6 m^3 · sec^{-1})
Pocatello	Red Rock Pass to Pocatello	---	106	95	0.85–1.15	1.0
Burley Basin	Raft River to Burley Basin	661–691	13	42	0.7–1.0	0.85
Murtaugh	Starrhs Ferry to Hansen Bridge	620–647	80	40	0.28–0.31‡	0.28
Eden Channel	Hynes Siding to north of Hansen Bridge	---	26	32	0.57–0.62‡	0.62
Hagerman	South of Jerome to King Hill	545–608	159	95	0.85–0.93	0.85
Swan Falls	Sinker Creek to Marsing	424–460	106	65	0.82–0.91	0.82
Farewell Bend	Porters Ferry to Dennett Creek	309–344	68	50	0.65–0.82	0.65
Pittsburg Landing	Bills Creek to Pittsburg Landing	214–234	90	32	0.57–0.62	0.57
Dug Bar	Pittsburg Landing to Salmon River	188–214	89	42	---§	0.57
Lewiston	China Garden to Lewiston	141–176	80	55	0.57–0.62	0.57

*Snake River Mileage as indicated on USGS topographic quadrangles. Estimated in locations where mileage is not labeled.
†Discharge estimate used for the calculation of local hydraulic conditions of depositional and erosional sites.
‡Discharge values for the Murtaugh and Eden channel reaches should be added together for comparison with the other reaches.
§Bracketing discharges were not determined for the Dug Bar reach. The utilized discharge is from consideration of estimates for adjacent reaches.

flow hydraulics, and (3) the desire to test the applicability of the step-backwater modeling in a variety of hydraulic and geologic environments.

The reaches varied in length from 32 to 95 km and were represented by 13 to 159 cross sections. All cross sections were digitized from 7.5-min USGS topographic quadrangles (except for portions of reaches within Hells Canyon where only 15-min maps were available). Cross sections were spaced an average of 0.67 km apart with distances ranging from 90 m to 5.7 km. In general, cross sections were spaced at intervals such that there were one to four cross sections in a channel length equivalent to the local flow width. Cross sections were located and oriented to represent the spatial variations in flow geometry and to be perpendicular to the presumed flow direction at maximum stages (Fig. 13). All reaches were modeled assuming subcritical flow conditions and the profiles were calculated in the upstream direction. For prescribing initial conditions for each modeled reach, estimates of the maximum stage (the starting stage for each model trial) at the downstream end of each reach were based on the local geologic evidence for maximum stages. Values of Mannings n and expansion-contraction loss coefficients (C) are difficult to assign for a flow of the scale of the Bonneville Flood because there are no actual discharge measurements for flows of such a large magnitude. Nevertheless, Mannings n values were assigned on a consistent basis (Table 2), and sensitivity tests indicated that uncertainties were of minimal importance relative to the modeled discharge value. Following Jarrett and Malde (1987), expansion and contraction coefficients of 0.5 and 0.0 were used for channel segments of expanding and contracting flow, respectively.

Study reaches and modeling results

The step-backwater calculations, in conjunction with the geologic evidence of maximum flow stages, provide precise discharge estimates (uncertainties are generally less than 25%) that are mutually consistent and indicate a progressive downstream

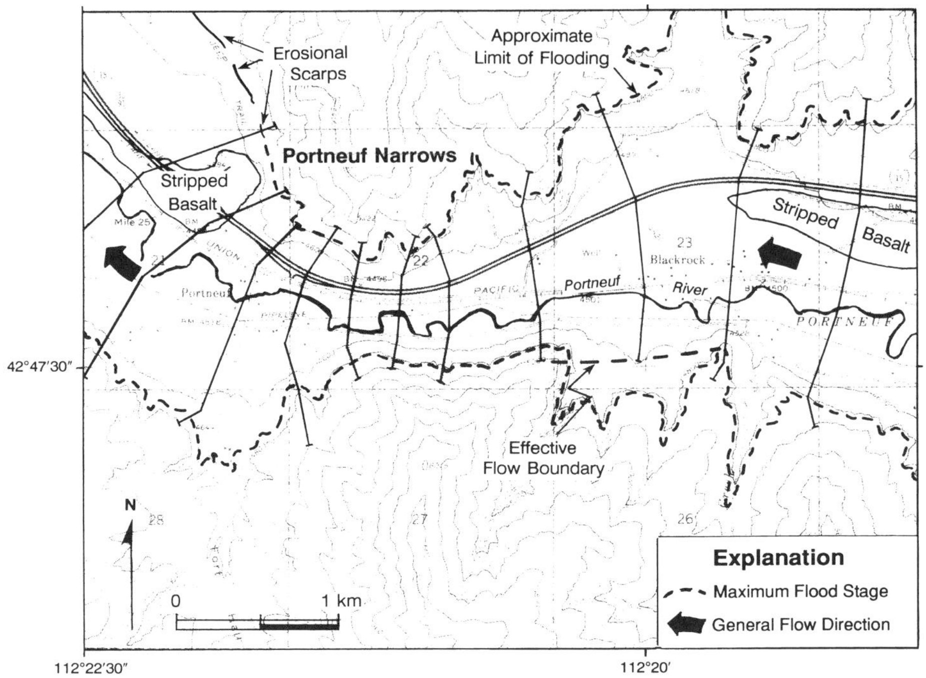

Figure 13. Topography and cross section locations within the Portneuf Narrows (Fig. 14). Basalt flows following the valley bottom were completely eroded away in the narrow part of the constrictions. Cross sections were placed and oriented to be perpendicular to the local flow direction. Only areas of effective flow (flow in the downstream direction) are included in the cross sections measured for the step-backwater calculations. Topography from the Inkom (ID) 7.5-min USGS Quadrangle.

TABLE 2. ASSIGNED MANNINGS *n* VALUES

Surface	Mannings *n* value
Main channel	0.03
Tractive deposits	0.03
Valley walls	0.05
Stripped basalt or bedrock uplands	0.05
Basin and butte scabland	0.06–0.1
Major flow obstacles	0.1–0.9

attenuation of peak discharge (Table 1). The best discharge estimates are for reaches that have constricted segments that experienced critical flow and where there is good evidence for maximum flood stages upstream of the constriction. Flow reconstructions in the upper reaches of the flood route are more tightly bracketed by the evidence of maximum flood stages.

Peak discharge at Red Rock Pass was $1 \pm 0.15 \times 10^6$ $m^3 \cdot sec^{-1}$. By Lewiston, at the downstream end of the studied portion of the flood route, peak discharge had decreased to 0.57 to 0.62×10^6 $m^3 \cdot sec^{-1}$. For several of the reaches, the calculated water-surface profiles mimic the geologic evidence of maximum stages fairly closely; this is reflected in the relatively high degree of precision that has been assigned to the discharge estimates (Table 1). Furthermore, the good match between the modeled and actual profiles for some of the reaches lends credibility to the analysis of the local hydraulics that is derived from the results of the flow reconstructions.

In the sections that follow, brief descriptions of flood features and summaries of the flow reconstructions are given for each of the modeled reaches. This is followed by general comments regarding the success in applying step-backwater modeling techniques for paleohydrologic analysis of the Bonneville Flood and the general nature of the hydrology and hydraulics of the flood.

Red Rock Pass to Pocatello. This reach encompasses the flood route from the area of release from Lake Bonneville near Red Rock Pass to its entrance onto the Snake River Plain near Pocatello (Fig. 14). Over this 95-km distance, the water-surface profile, at maximum stages, dropped from the 1,552-m (5092 ft) Lake Bonneville level at Red Rock Pass to 1,378 to 1,382 m (4,520 to 4,535 ft) a few km north of Pocatello (Fig. 15). Much of the water-surface drop was through prominent constrictions; the major ones were where the flood cut through the Marsh Creek fan near Red Rock Pass (Fig. 6), at the Portneuf Narrows

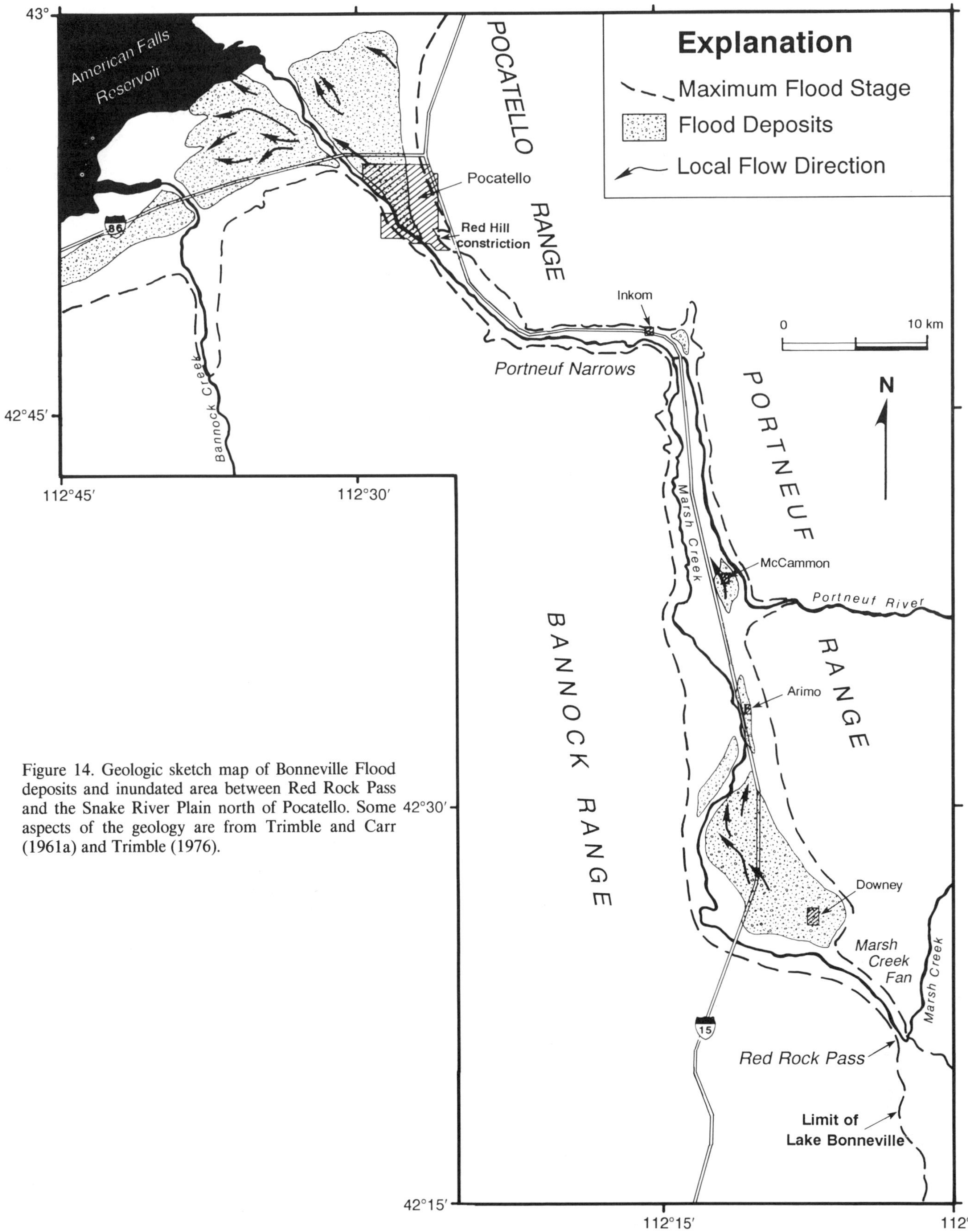

Figure 14. Geologic sketch map of Bonneville Flood deposits and inundated area between Red Rock Pass and the Snake River Plain north of Pocatello. Some aspects of the geology are from Trimble and Carr (1961a) and Trimble (1976).

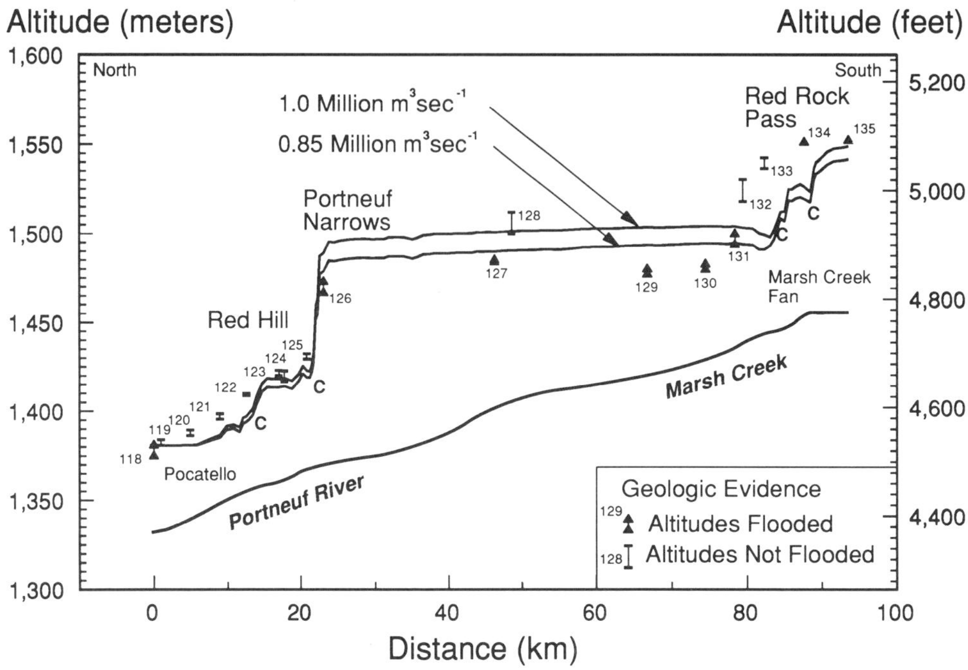

Figure 15. Geologic evidence for maximum flood stages and modeled water-surface profiles for the Pocatello reach. For all reach profiles (Figs. 20, 24, 25, 30, 39, 41, 45, 46, 49), the estimated limits of uncertainty in determining the altitudes of evidence for maximum flood stages are given by the upper and lower symbols for each numbered stage indicator. The numbers refer to descriptions of the feature provided in Appendix 2. Locations that are calculated as having critical flow are labeled with the letter C.

(Fig. 13), and at Red Hill. For the remainder of the reach, the high-water indicators (Fig. 15) indicate that the flood profile was fairly flat. With the exception of the constrictions at Red Rock Pass, Red Hill, and the Portneuf Narrows, the margins of the flood were generally confined by erosional scarps developed in the extensive piedmont deposits along the flanks of the Bannock and Portneuf Ranges. In wider reaches there are extensive flood deposits of reworked piedmont gravel (Fig. 16). The most spectacular of these is downstream of Pocatello where a huge expansion bar was deposited where the flood entered the Snake River Plain (Trimble and Carr, 1961a; Trimble, 1976). In narrower reaches, there was intense erosion. For example, through the Portneuf Narrows, the flood removed portions of two basalt flows that follow the valley bottom for much of the northern part of this reach (Figs. 7, 17).

A reconstructed water-surface profile was developed from step-backwater calculations through 105 cross sections along the length of the reach. The water-surface profile associated with a 1.0×10^6 $m^3 \cdot sec^{-1}$ discharge is consistent with the geologic evidence of maximum stages along the flood route and with the maximum altitude of the surface of Lake Bonneville. A calculated profile associated with discharge of 0.85×10^6 $m^3 \cdot sec^{-1}$ is the minimum discharge that could have emplaced the majority of high-water marks along the flood route (Fig. 15). Flow is modeled as having achieved critical or supercritical conditions for short reaches (one to four cross sections) at the three major constrictions described above and for a single cross section at Red Rock Pass. Between constrictions the calculated water-surface profiles have much lower gradients in a manner that is consistent with the geologic evidence.

Achievement of critical flow at Red Rock Pass provides the only opportunity along the flood route to determine a *maximum* value for the peak discharge of the Bonneville Flood. For most of the flood reach it is much easier to establish minimum peak discharge estimates because of the greater degree of certainty associated with establishing minimum estimates of altitudes of maximum stages. However, upstream of Red Rock Pass, the maximum Lake Bonneville level has been well established at 1,552 m (5,092 ft) (Currey, 1990); this altitude represents the maximum water-surface and energy level that can be associated with the flood profile. The attainment of critical flow (minimum specific energy) at Red Rock Pass indicates that once the present topographic configuration was established at the outlet, the rate of outflow from Lake Bonneville was controlled by the rate of flow through this bedrock constriction. Furthermore, the enlargement of the cross sections at Red Rock Pass and at Marsh Creek during the flood would have led to continually increasing flow conveyance as the flood progressed; this implies that the present topography closely represents *maximum* potential flow conveyance through the breakout area (less the minimal deposi-

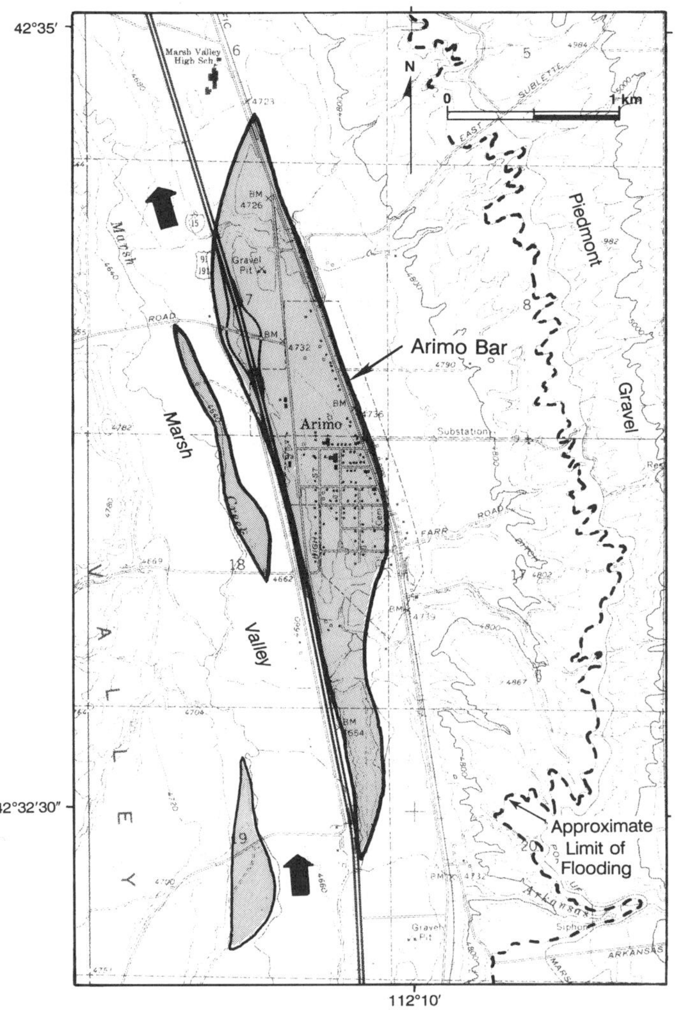

Figure 16. Longitudinal bars near Arimo within the Marsh Creek Valley (Fig. 14). The bars are primarily composed of reworked piedmont gravel that flanks the margins of the flood route downstream of Red Rock Pass. The mapped limits of flooding are interpolated from evidence of maximum stages upstream and downstream of this location. Topography from the Arimo (ID) 7.5-min USGS Quadrangle.

Figure 17. Scoured and fluted basalt surface near the Portneuf Narrows (Fig. 14). Flow was from left to right across the surface. G. K. Gilbert, in his field notes of September-October 1879 (in Hunt, 1982, p. 105–118), described a similar surface nearby: "I find the surface broadly waterworn—large flutings and potholes."

tion that has occurred subsequent to the flood). Hence, the maximum possible discharge that could have issued from Lake Bonneville is one that achieves critical flow at the constriction and has an associated water-surface profile that achieves a maximum upstream flow stage that equals the 1,552-m (5,092 ft) Lake Bonneville level (Fig. 18). For there to have been a larger discharge through this constriction, either the cross-sectional area at the constriction must have been larger than present, or the maximum Lake Bonneville level was higher than presently thought. Neither of these possibilities is likely. Therefore, an upper limit for the maximum discharge of the Bonneville Flood can be firmly established at about 1.1×10^6 $m^3 \cdot sec^{-1}$. This value is close to the minimum estimate established by consideration of the high-water indicators along the length of the reach, lending credibility to the general accuracy of the discharge estimates determined by consideration of minimum estimates of maximum flood stages.

Raft River to Burley Basin. The Burley Basin reach is a wide, low-gradient portion of the flood route (Fig. 19). Upstream of this reach is the Massacre Rocks area where there were substantial and complex channel changes during the flood associated with the breaching of the basalt dam that impounded Pleistocene American Falls Lake (Trimble and Carr, 1961b; Scott and others, 1982). Downstream of the Burley Basin is the one section of the flood route where two separate flow paths were followed as the flood spilled out of the basin at two locations.

At the upstream end of the reach, near the Raft River confluence, erosional scarps in loess and the distribution of gravel on the basalt uplands indicate that maximum flood stage was 1,308 to 1,311 m (4,290 to 4,300 ft). Maximum stage at the downstream end of the reach is defined at the two outlets of the Burley Basin where erosional features on basalt indicate maximum stages

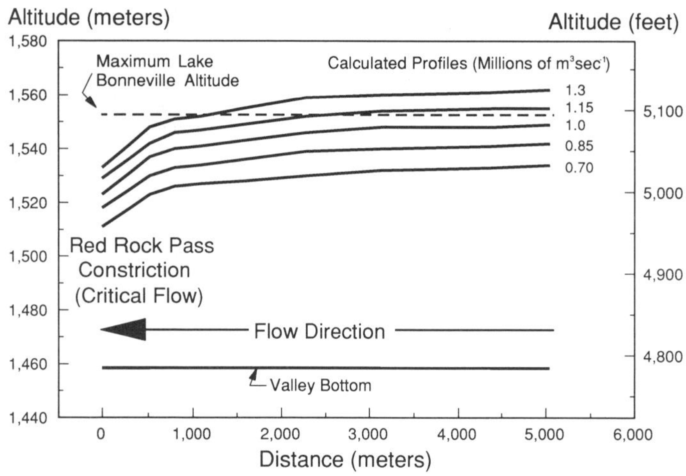

Figure 18. Step-backwater profiles through the Red Rock Pass constriction for discharges between 0.7 and 1.3 million $m^3{\cdot}sec^{-1}$. These profiles were calculated for conditions of critical flow at the Red Rock Pass constriction. About 1.10 million $m^3{\cdot}sec^{-1}$ is the maximum discharge that could have been conveyed through the Red Rock Pass constriction associated with the maximum level of Lake Bonneville and the present topography.

of 1,295 to 1,298 m (4,250 to 4,260 ft) (Fig. 20). The extremely low gradient of the flood (the water surface descended only 10 to 16 m over 42 km) is reflected by the deposits in this reach. Only in the upstream end of the reach, where energetic flow was entering from the Massacre Rocks area, are there large accumulations of flood deposits. Downstream, in the Burley Basin, tractive deposits are restricted to cobble bars emplaced near the present course of the Snake River, and sand and silt farther away from the main channel. Extensive accumulations of slackwater sediment were deposited as high as 1,286 m (4,220 ft) along the north margin of the basin.

Flow was modeled through 13 cross sections over a distance of 42 km. The relatively long interval between cross sections is because of the extreme width (10 to 30 km) of the inundated area. Water-surface profiles associated with discharges of 0.7 to 1.0×10^6 $m^3{\cdot}sec^{-1}$ are consistent with the evidence of maximum flood stage (Fig. 20). Because of the low water-surface gradient and the absence of a well-defined profile provided by the geologic evidence, the modeling results are consistent with a wide range of possible discharges. The reconstructed profiles are flat for the downstream 20 km, indicative of the hydraulic ponding behind the two outlets of the Burley basin.

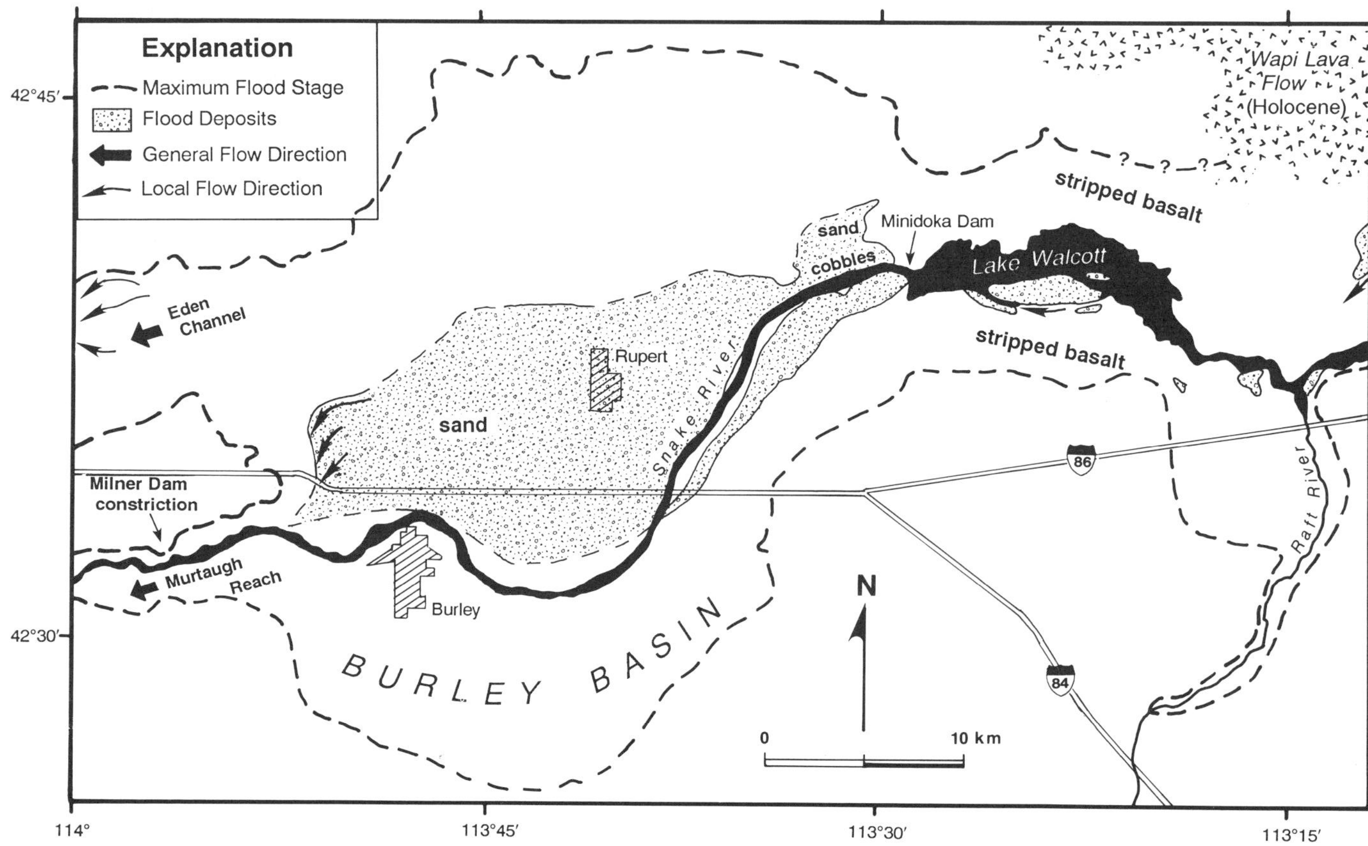

Figure 19. Map of Bonneville Flood features and inundated area between the Raft River confluence with the Snake River and where the flood split into two paths downstream of the Burley Basin. The limits of the inundated area were mapped on the basis of the upper extent of flood deposits and stripped basalt in conjunction with the basin topography.

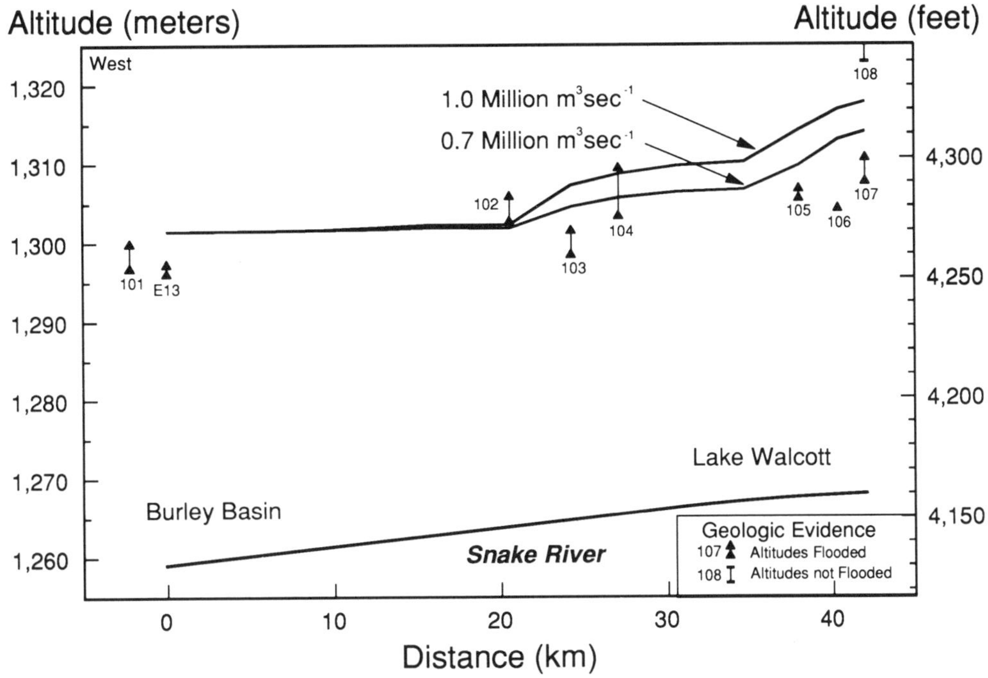

Figure 20. Reconstructed water-surface profiles and geologic evidence for maximum flood stages for the Burley Basin reach. The downstream constraints on the maximum flood stage are given by evidence of the maximum flood height at the outlets of the basin.

Murtaugh reach. At stages above 1,279 m (4,195 ft), flow exited the Burley Basin by two routes; one path, the Murtaugh reach, is the present course of the Snake River. The other, the Eden channel, was over the broad basaltic uplands north of the river (Fig. 21). The two flow routes rejoined near physiographic Twin Falls where flow from the Eden channel spilled back into the Snake River canyon and eroded the spectacular cataracts that incise the north rim of the canyon over a 15-km-wide swath. The modeled Murtaugh reach encompasses the Snake River canyon from 7 km west of Burley to the Hansen Bridge, 5.5 km upstream of physiographic Twin Falls. Malde's (1968, p. 23–25) description of the flood features in this reach has been partially drawn from here.

The Murtaugh reach is a very different hydraulic environment than that of the Raft River to Burley reach. It is the steepest section of the flood route, following a narrow, deeply incised canyon. The river flows on basalt for much of its length (Figs. 22, 23). Both the steepness and narrowness of the canyon result from the recent southward diversion of the Snake River to its present position by late Quaternary flows of the Sand Springs Basalt that filled an earlier canyon (Malde, 1971; Covington and others, 1985).

The amount of flow entering the Murtaugh reach during the Bonneville Flood was regulated by a constriction 5 km upstream of Milner Dam (Fig. 21) that hydraulically impounded water in the Burley Basin to an altitude of 1,293 to 1,297 m (4,250 to 4,260 ft). At the downstream end of the reach, 1 km west of Hansen Bridge, where the two separate flood routes merge, upland channels and erosional scarps indicate a maximum stage of at least 1,189 m (3,900 ft). Much of the intervening flood profile is defined by erosional features along the margins of the canyon (Figs. 23, 24). The canyon was brimful with flow spilling over and onto the adjacent uplands for much of its length. Two kilometers downstream of the present location of Milner Dam, a small portion of the flow crossed the divide and spilled north into the Eden channel (Scott, 1982).

Flood features are primarily erosional because of the narrowness of the canyon and the steep gradient. Extensive tracts of butte-and-basin scabland and stripped basalt flank the Snake River in wider reaches of the canyon and on the uplands where substantial flow spilled out of the canyon. In narrow canyon sections, there is little evidence of the flood, only a rectangular cross section swept clean of talus and a basalt channel bottom (Fig. 22). Flood deposits are confined to isolated locations on the uplands, a few small bars that flank the river in wider sections of the canyon, and in reentrants within the canyon walls. Malde (1968, p. 23) suggested that the position of the base of one of these bars 15 m above the present level of the Snake River is indicative of 15 m of postflood downcutting. Nearby, however, Bonneville Flood deposits are preserved at the present river level, indicating that there has been little downcutting of the Snake River subsequent to the flood.

Flow was modeled through 80 cross sections for a distance of 40 km. A minimum discharge of 0.28×10^6 $m^3{\cdot}sec^{-1}$ is

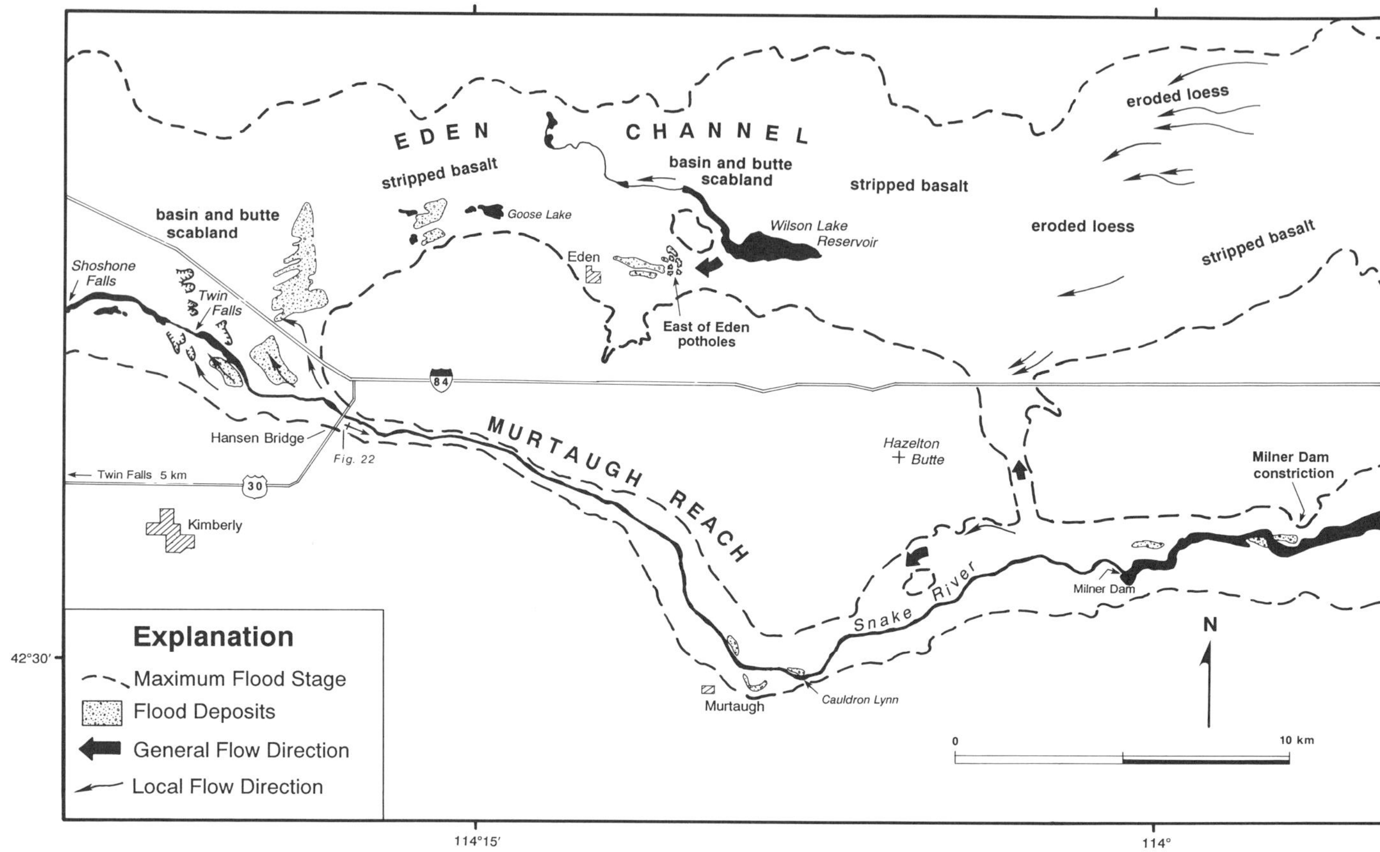

Figure 21. Bonneville Flood features and inundated area for the Eden channel and Murtaugh reaches. Maximum flood stages are primarily constrained by the upper extent of erosional features. At the west end of this area, several cataracts formed where flow from the Eden channel spilled back into the Snake River Canyon near Twin Falls and Shoshone Falls.

consistent with the geologic evidence of maximum ponding levels in the Burley Basin above the constriction at Milner Dam (Fig. 24). For the remainder of the reach, the modeled profile provides a poor fit to the geologic data. As is discussed in more detail in the section summarizing the quality of the modeling results, the poor fit is probably the result of unsatisfactory accounting of energy losses in the step-backwater calculations.

The calculated profile is stepped, with flow passing through critical depth at four locations. The constricted reach near Hansen Bridge had a calculated water-surface drop of 60 m and flow velocities that exceeded 30 m $\cdot$ sec^{-1}. Another important constriction in this reach is at the upstream end of the reach, near Milner Dam. Although the channel here is several times wider than at constrictions downstream, the Milner Dam constriction regulated flow exiting the Burley Basin and forced ponding to an altitude that forced a major portion of the flood, at high stages, to exit the basin through the Eden channel.

Eden channel. Malde (1968, p. 22–23) first described this reach and assigned the name "Rupert channel." It seems more geographically appropriate, however, to call it the Eden channel. The Eden channel closely follows the surface of the Sand Springs Basalt, a late Quaternary, 80-km-long basalt flow that filled a preexisting canyon of the Snake River and diverted the river south to its present position in the Murtaugh reach. The Eden channel was unique along the flood route in that it is the only reach of substantial flow that did not follow a previously existing watercourse.

Water entering the Eden channel spilled out of the Burley Basin through a broad, 7.5-km-wide swale 16 km northwest of Burley (Fig. 21). Stripped basalt surfaces and rounded nonbasalt pebbles at the margins of the outlet indicate a maximum stage of at least 1,297 m (4,255 ft) where the flood exited the Burley Basin. The lowest point in the swale is at 1,279 m (4,195 ft), indicating there was a maximum water depth of 22 m at the outlet and that this reach only conveyed flow while there was ponding to altitudes greater than 1,279 m (4,195 ft) in the Burley Basin.

The flood route from the outlet to where it reentered the Snake River canyon near the city of Twin Falls was steep and wide. Flow width was on the order of 10 km and the maximum

Figure 22. View east (upstream) of the Snake River Canyon within the Murtaugh reach about 0.5 km upstream of the Hansen Bridge (photograph location shown in Fig. 21). The canyon is 100 m deep and floored by basalt. The river at low flow is composed of a series of large pools separated by narrow slots cut in the basalt. At maximum Bonneville Flood stage, flow filled the canyon and spilled onto the adjacent uplands.

Figure 23. Oblique aerial view to the east (upstream) of the Snake River Plain and the incised Snake River within the Murtaugh reach. The maximum extent of erosion (stripped basalt surfaces) by the flood is shown fairly precisely by the limits of cultivation on the uplands. The large set of falls in the middle foreground is Cauldron Lynn (Fig. 21), which stopped the Hunt Overland Party's 1811 attempt to descend the Snake River in wooden canoes.

stage dropped from 1,297 m (4,255 ft) at the outlet to 1,180 m (3,870 ft) over a distance of 40 km (Fig. 25). Abundant erosional features define the intervening flood profile. In many areas, the loess cover is extensively channeled and shaped into streamlined forms; in areas of higher erosional intensity, the loess cover has been completely removed, leaving stripped basalt surfaces. In reaches of greatest erosional intensity, the basalt has been eroded into butte-and-basin scabland topography. The most spectacular features are the 15- to 40-m-deep, 200-m-wide scour holes, here called the "East of Eden potholes," excavated into the basalt surface (Malde, 1968, p. 22) (Figs. 26, 27). Flood deposits are scarce in this reach, restricted to two small groups of bars at the downstream end of the reach and a series of bars that emanate from the potholes.

Step-backwater profiles were calculated through 26 cross sections over a distance of 38 km. Discharges of 0.57 to 0.62 $\times 10^6$ $m^3 \cdot sec^{-1}$ are consistent with the geologic evidence of maximum stages for the upstream part of the reach and in the Burley Basin (Fig. 25). Downstream of the East of Eden potholes, the water-surface profile was steeper, and, similar to the Murtaugh reach, the calculated water-surface profiles are higher than indicated by the geologic evidence.

A discharge estimate of 0.57 to 0.62 $\times 10^6$ $m^3 \cdot sec^{-1}$ for the Eden channel, added to the discharge estimate for the Murtaugh reach (0.28 $\times 10^6$ $m^3 \cdot sec^{-1}$), yields a total discharge through this section of the flood route of 0.85 to 0.90 $\times 10^6$ $m^3 \cdot sec^{-1}$. This is consistent with discharge estimates obtained in upstream as well as downstream reaches (Table 1). Considering these discharge estimates, 60 to 70% of the total flow followed the Eden channel at peak discharge.

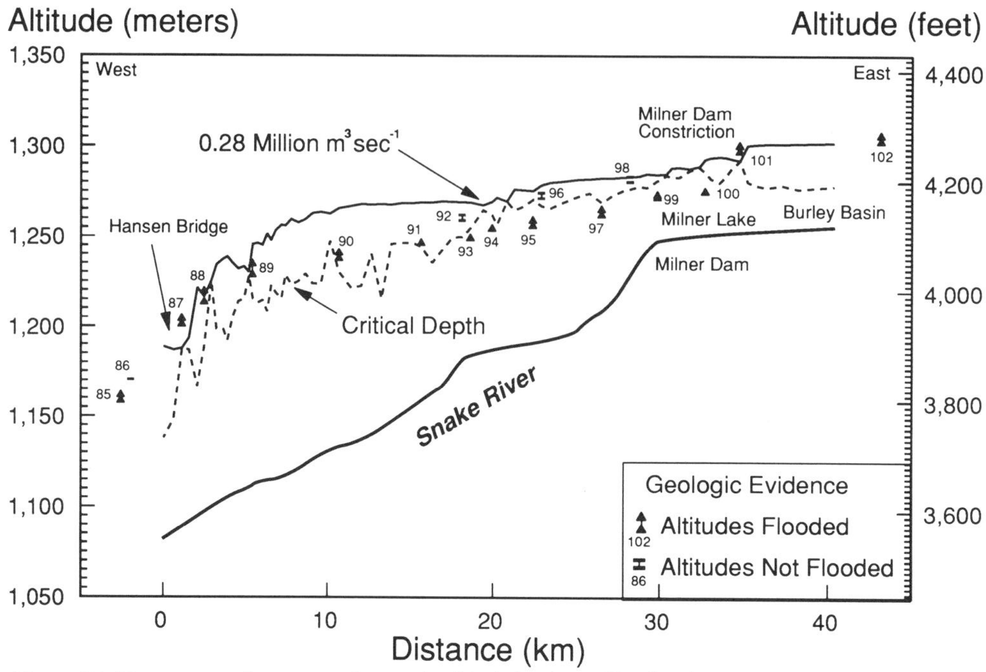

Figure 24. Reconstructed water surface and critical depth profiles for the Murtaugh reach for a 0.28 million $m^3{\cdot}sec^{-1}$ discharge. Flow probably was critical at the Milner Dam constriction, and the discharge estimate of 0.28 million $m^3{\cdot}sec^{-1}$ was assigned on the basis of maximum stage evidence upstream of the constriction. The calculated water-surface profile generally provides a poor fit to the geologic evidence downstream of the Milner Dam constriction where the highwater evidence suggests flow was at, or near, critical depth for substantial distances. Possible reasons are discussed in the text.

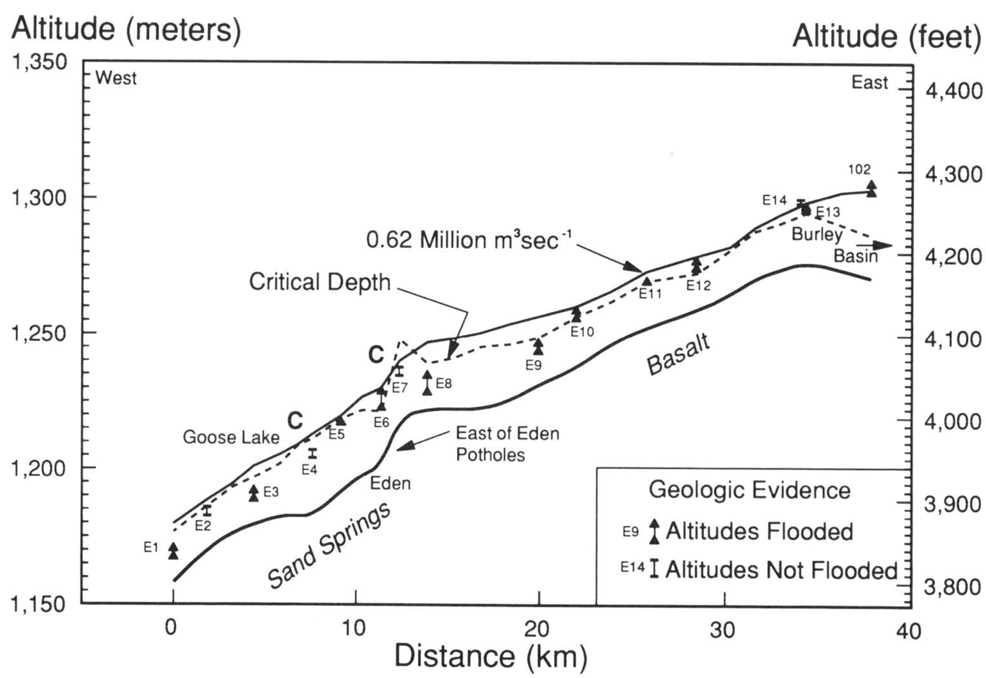

Figure 25. Geologic evidence of maximum flood stages and calculated water-surface and critical depth profiles for a 0.62-million $m^3{\cdot}sec^{-1}$ discharge within the Eden channel. The profile for this discharge matches the geologic evidence of maximum stages upstream of the sections of critical flow and results in a calculated ponding level in the Burley Basin consistent with evidence of maximum stages at the two outlets. Similar to the Murtaugh reach, a substantial length of the channel apparently had flow, at or near critical depth.

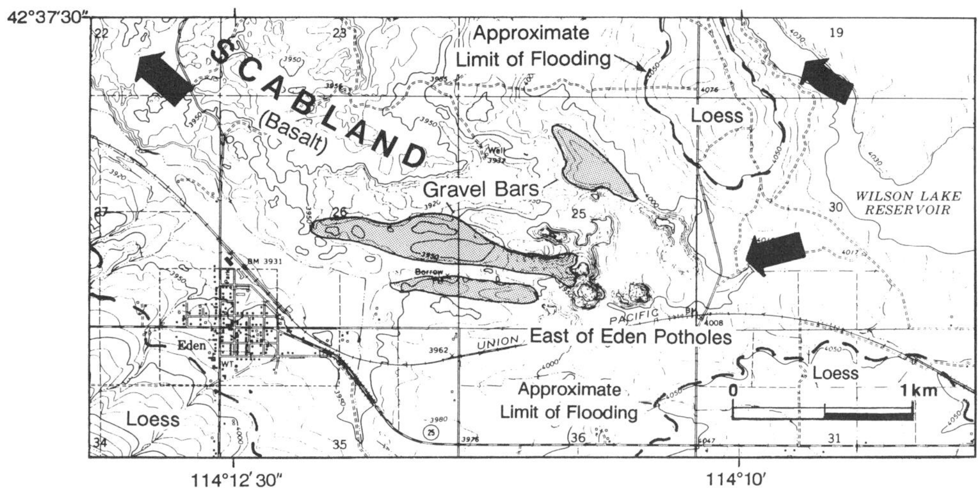

Figure 26. Topography and inundated area near the East of Eden potholes within the Eden channel (Fig. 21). Some of the "potholes" have depths exceeding 30 m and may have resulted from erosion by cavitation (discussed in text). Adapted from Malde (1968, p. 22–23). Topography from the Eden (ID) 7.5-min USGS Quadrangle.

Figure 27. Southeast directed (upstream) view of one of the deeper East of Eden potholes (Fig. 26). The photograph was taken from a boulder accumulation that emanates downstream from the scoured depression. The bottom of this feature is about 40 m below the basalt surface. Note people for scale.

The flow from the Eden channel reentered the Snake River canyon and joined the flow from the Murtaugh reach by flowing over the north rim of the canyon in a broad 15 km swath between physiographic Twin Falls and a position due south of Jerome. The complicated hydraulics of this short reach defy reconstruction of the flow conditions; nevertheless, this section is noteworthy because it contains some of the most spectacular erosional features of the flood (Malde, 1968, p. 25–30). As the flow spilled into the canyon, it carved several large cataracts (Fig. 11) that eroded as far as 2 km headward. Flow within the canyon dropped over several large falls, including the 65-m-high Shoshone Falls (Malde, 1987) and the 45-m-high Twin Falls, and carved large scabland complexes in their vicinity. South of Jerome, the flow on the uplands and in the canyon completely rejoined and was primarily confined within the Snake River canyon as it entered the Hagerman reach.

Hagerman reach. The Hagerman reach extends from south of Jerome to the town of King Hill (Figs. 3, 28). Along with the reach at Swan Falls, it is one of the important reaches in evaluating boulder transport and deposition. Malde (1968, p. 30–37) provided a detailed description of many of the flood features within this reach.

The Hagerman reach comprises a series of narrow basalt-walled canyon sections separated by basins formed where one or both of the valley margins are underlain by Quaternary alluvial deposits. The canyon segments were generally brimful with water spilling out onto the upland margins at isolated locations. The bottoms of the canyons were eroded by the flood to a scabland topography, although, in wider sections, there are some longitudinal boulder bars. The major basins of Melon Valley (Fig. 29), Hagerman Valley and King Hill basin (Fig. 28), were sites of substantial deposition of material eroded from the constricted

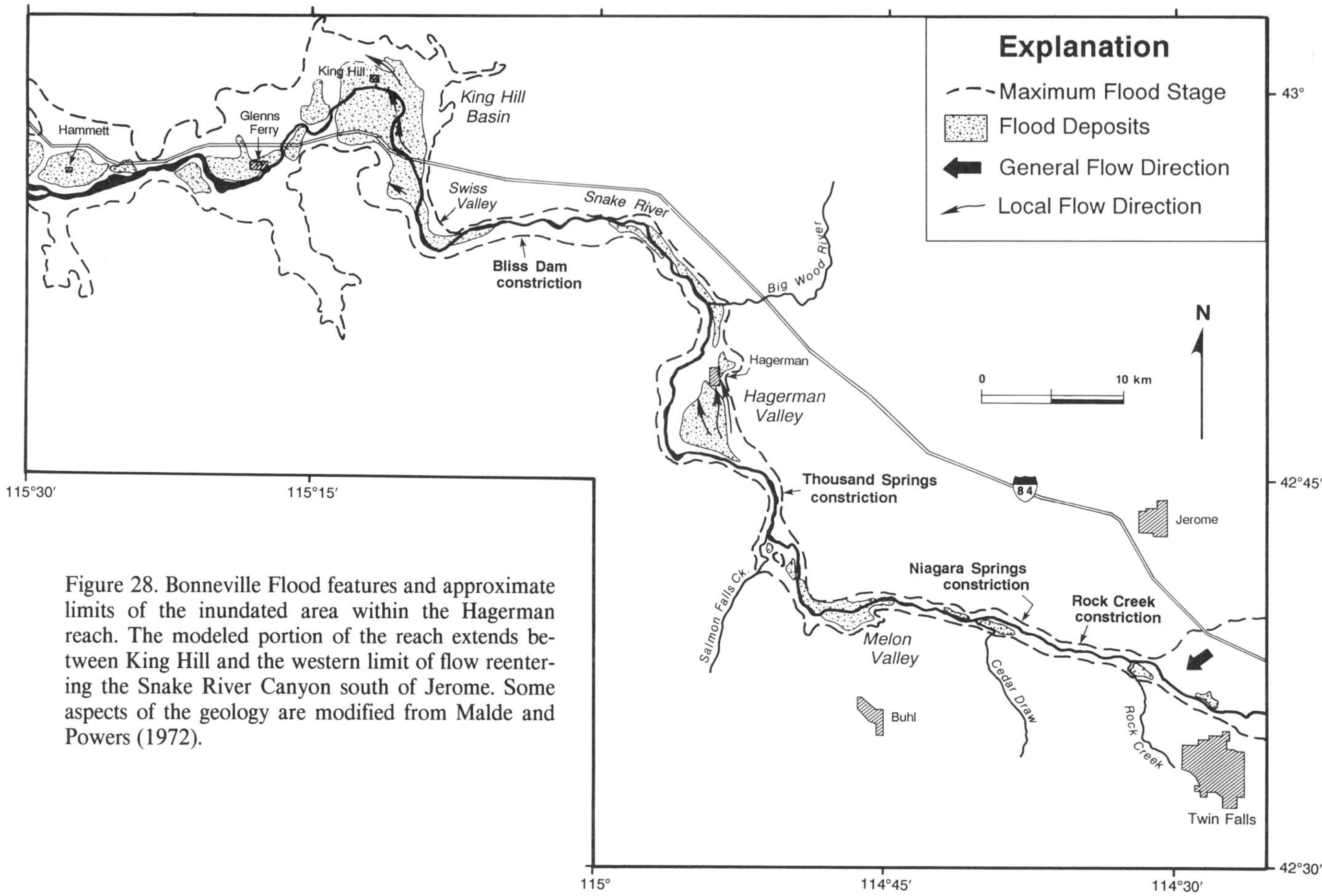

Figure 28. Bonneville Flood features and approximate limits of the inundated area within the Hagerman reach. The modeled portion of the reach extends between King Hill and the western limit of flow reentering the Snake River Canyon south of Jerome. Some aspects of the geology are modified from Malde and Powers (1972).

sections. The expansion bar at the south end of Hagerman Valley is 5 km long and mantled with boulders, some with long diameters exceeding 4 m.

The flood profile was steep through this reach, dropping from a maximum stage of at least 1,119 m (3,570 ft) at the upstream end to an altitude of 869 to 876 m (2,850 to 2,875 ft) downstream at King Hill. The intervening profile is fairly well defined by deposits and erosional features on the adjacent uplands (Fig. 30). Step-backwater calculations were computed through 159 cross sections over a distance of 92 km. A discharge of 0.85×10^6 $m^3 \cdot sec^{-1}$ gives a calculated water-surface profile consistent with most of the geologic evidence for maximum stages and is also consistent with convincing evidence of maximum flood stages at the upstream end of the reach (Fig. 30). Similar to many of the modeled water-surface profiles in steep reaches of the flood route, the modeled profile drops in a series of steps that are associated with constricted segments. The steep drops are separated by ponded areas or reaches of lower water-surface gradient.

One of the more notable constrictions along the entire flood route was near Rock Creek (Figs. 8, 31). Here the flow is calculated as achieving critical or supercritical conditions for several cross sections and having a water-surface profile that dropped 100 m over a distance of 3 km. At the Rock Creek constriction, calculated maximum flow velocities exceeded 35 $m \cdot sec^{-1}$. Other constrictions in this reach that probably experienced critical flow are near Bliss Dam and the short reach between Melon and Hagerman Valleys (Fig. 30).

Immediately downstream of both the Rock Creek and Bliss Dam constrictions, basalt surfaces stripped of loess that flank the canyons indicate maximum flood stages higher than those predicted by the step-backwater calculations (Figs. 30–32). This could reflect the inability to model adequately the flow through these constrictions. However, these discrepancies more likely relate to changing channel geometry during the flood. For the Rock Creek constriction, it is plausible that, prior to the flood, the segment of canyon immediately downstream of the present constriction was much narrower than present. The downstream limit of the constriction early in the flood may be indicated by the most upstream limit of flood deposits on the canyon floor and the downstream limit of anomalously high flood features on the uplands (Fig. 30). As the flood progressed, erosion would have been most intense at the downstream end of the constricted reach where most of the water-surface drop and energy loss occurs.

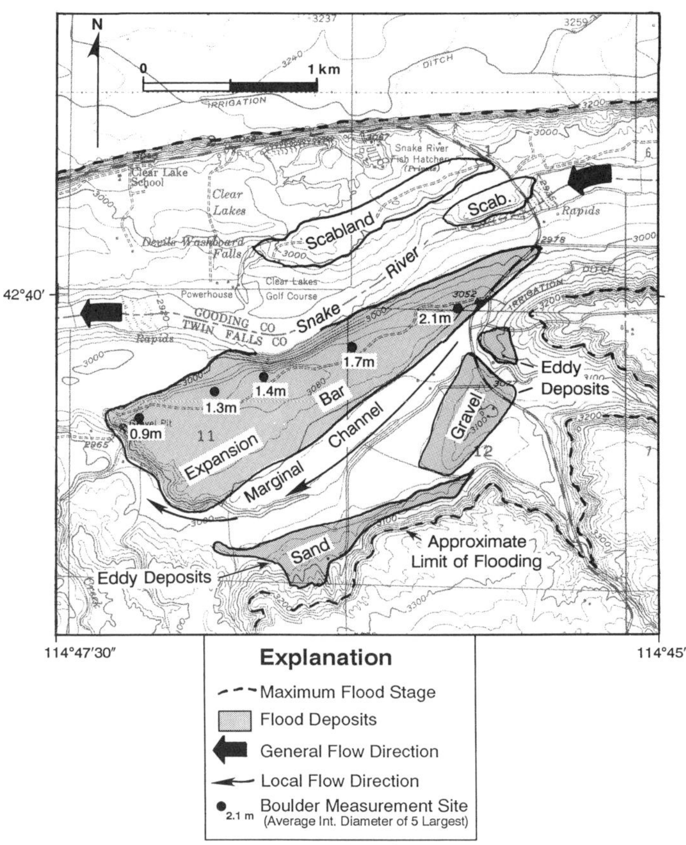

Figure 29. Some key flood features and topography within a portion of Melon Valley (Fig. 28). The forms and locations of the expansion bar and eddy bar are typical of many such expansions along the flood route. Note that, for clarity, not all Bonneville Flood deposits are shown. Topography from the Thousand Springs (ID) 7.5-min USGS Quadrangle.

Consequently, the constriction would have enlarged, and this widening would have progressed upstream, perhaps almost doubling the conveyance of the canyon in the present area of extensive scabland topography between the past and present downstream limits of the constriction (Fig. 31).

In a similar manner, it is likely that the canyon segment near Swiss Valley (Figs. 28, 33) was much narrower early in the flood, and, consequently, water flowed over the adjacent basalt uplands. As flow eroded into the soft fluvial and lacustrine sediment of the Glenns Ferry Formation along the outside of the bend (the large point bar at Swiss Valley is indicative of flow migration), the position of the constriction and its associated high ponding levels behind it migrated upstream to near the location of Bliss Dam. The maximum altitudes of flood deposits within the canyon are probably more faithful recorders of maximum stages once steady flow conditions were achieved and do provide a better match to the calculated water-surface profile (Fig. 30).

These two sites are the only ones along the flood route where there is sufficient evidence for the water-surface profile and geomorphic evidence of erosion and deposition for proposing scenarios such as the above. However, many of the anomalously high maximum flood-stage indicators along the flood route are located downstream of constrictions. The altitudes of these features may indeed have resulted from flow stages caused by a flow geometry different than one that would result from the present topography.

Downstream of the Hagerman reach, between King Hill and Swan Falls, the flood profile was remarkably flat, dropping only 30 to 45 m over a distance of 155 km, with a large portion of that drop occurring near Crane Falls (Fig. 10). This reach is nearly devoid of flood features, except for fine-grained deposits downstream of King Hill and erosional tracts and deposits left by energetic flow conditions near Crane Falls (Malde, 1968, p. 37–42). The water-surface profile is only tenuously defined by subtle erosional features and the altitude of flood features at the upstream end of the Swan Falls reach.

Swan Falls to Marsing. The Swan Falls reach extends from the Swan Falls constriction, upstream of Swan Falls, to the town of Marsing (Fig. 34). This reach is the most downstream portion of the flood route studied by Malde (1968, p. 42–44) and, with high-water evidence near the Swan Falls constriction, is where previous discharge estimates for the Bonneville Flood have been calculated (Clifford Jenkins, in Malde, 1968, p. 11–12; Jarrett and Malde, 1987). This reach of the Snake River encompasses the morphologic transition between an incised canyon within Snake River Plain basalt flows and the wide, low-gradient alluvial basin that the river follows before turning north and entering the gorges along the Oregon-Idaho border.

Within the Swan Falls constriction and the narrower canyon segments there are no flood deposits and the canyon walls are devoid of talus. However, in wider reaches, there are extensive longitudinal and expansion bars, the bars immediately downstream of the constriction (the bars near Priest Ranch) have abundant well-rounded boulders exposed on the surface with maximum diameters exceeding 10 m (Fig. 35). Farther downstream, the Initial Point and Walters Butte bars (Figs. 36–38) are large deposits of material eroded from the narrower canyon segments. At the downstream end of the reach, flow was less energetic and primarily deposited sand and gravel.

Jarrett and Malde (1987) reported an important high-level flood deposit immediately upstream of the Swan Falls constriction. This small bar, at an altitude of 829 m (2,719 ft), and a nearby eroded basalt surface with a maximum altitude of 840 m (2,755 ft), provide minimum constraints for the maximum flood stage upstream of the constriction. Downstream evidence indicates that the profile dropped quickly through the constriction and the narrow canyon segments below, to an altitude of 783 to 796 m (2,570 to 2,590 ft) near the upstream end of Walters bar and to 762 to 768 m (2,500–2,520 ft) at the downstream end of the reach (Fig. 39).

Flow profiles were calculated through 106 cross sections representing the 60 km reach. A minimum discharge of 0.82 to 0.85×10^6 $m^3 \cdot sec^{-1}$ gives a calculated profile consistent with

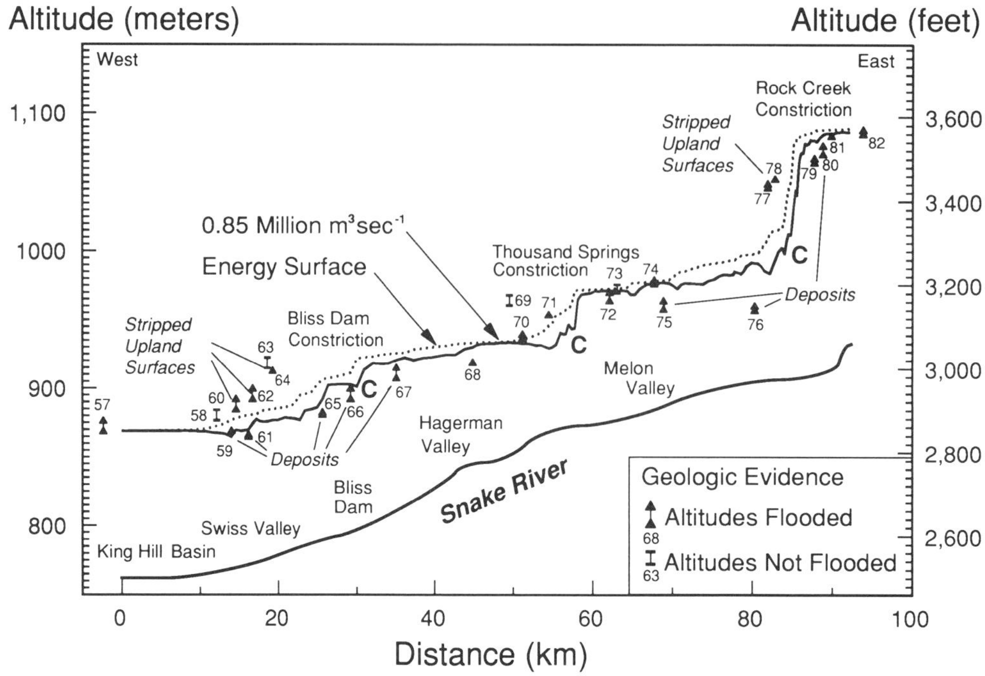

Figure 30. Geologic evidence of maximum flood stages and calculated water-surface and energy-surface profiles for a 0.85-million $m^3 \cdot sec^{-1}$ discharge within the Hagerman reach.

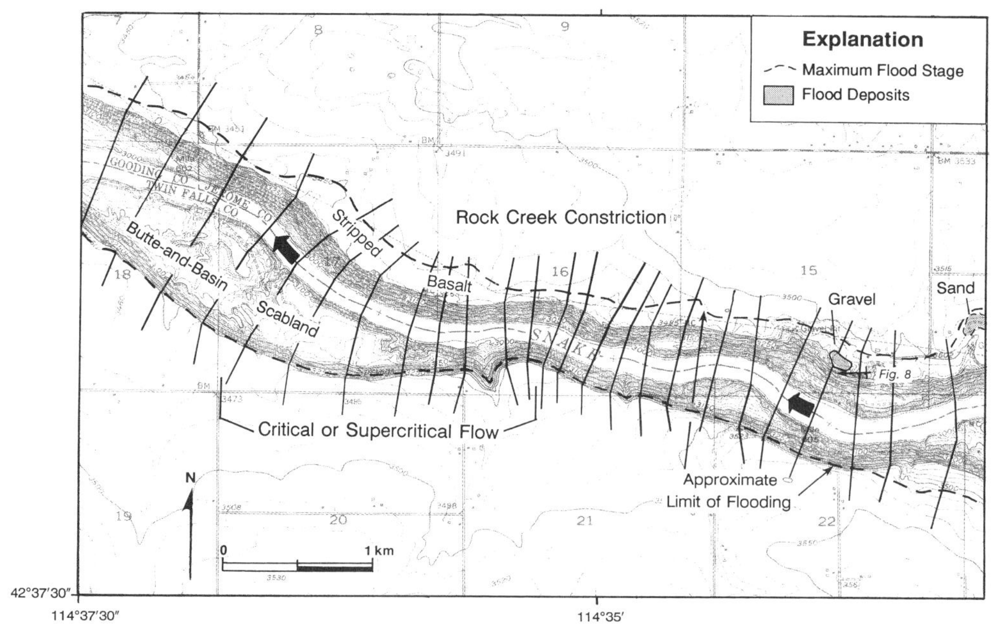

Figure 31. Bonneville Flood features, topography, and measured cross-section locations at the Rock Creek constriction (Figs. 8, 28). This was one of the most constricted reaches along the entire flood route; the entire discharge was funneled through a canyon 400 m wide by about 150 m deep. Several cross sections within the constriction are calculated to have experienced critical or supercritical flow. Limits of the flood were estimated from the upper extent of stripped basalt and flood deposits along the northern margin of the canyon. Topography from the Jerome 7.5-min (ID) USGS Quadrangle.

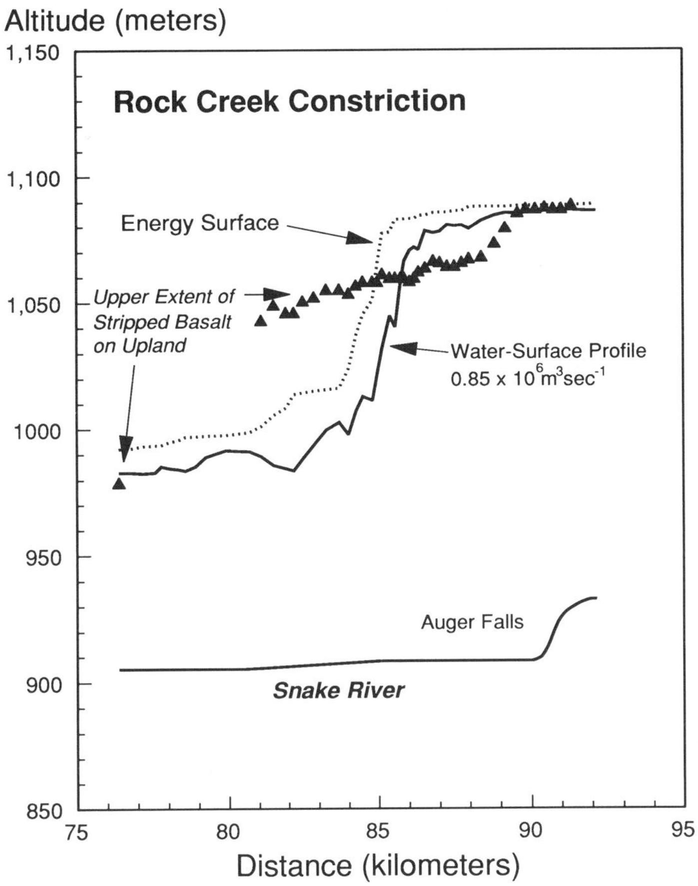

Figure 32. Calculated water-surface and energy profiles for a 0.85-million $m^3 \cdot sec^{-1}$ discharge through the Rock Creek constriction (Fig. 28). Maximum flood stages are defined by the upper limit of stripped basalt along the margins of the canyon. The distance values on the abscissa refer to the distance from the downstream end of the modeled Hagerman reach (Fig. 30).

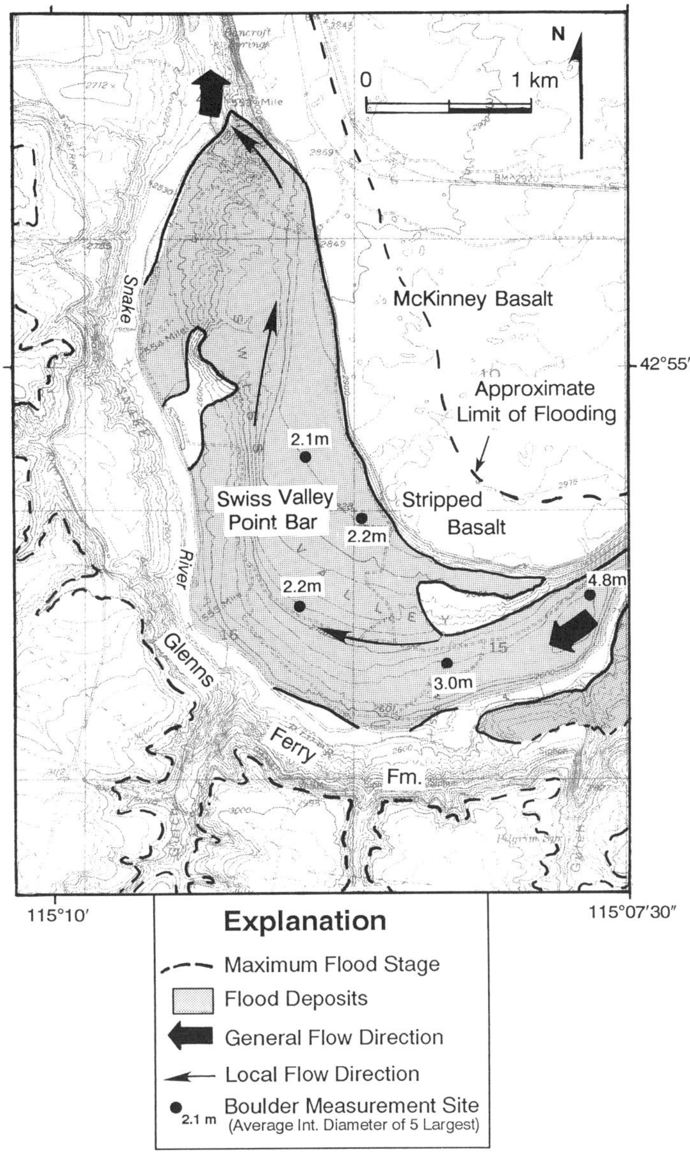

Figure 33. Flood features and topography in the vicinity of the Swiss Valley point bar. As discussed in the text, flow on the basalt upland north and east of the Snake River may have preceded canyon widening along the outside of the bend and development of the point bar during the flood. Topographic base from the Pasadena Valley (ID) 7.5-min USGS Quadrangle.

most of the geologic evidence of maximum stages over the length of the reach, and, most importantly, results in an appropriate water-surface altitude upstream of the Swan Fall constriction with respect to the flood features identified by Jarrett and Malde (1987). The calculated profile is stepped through the upper part of the reach with the three major drops all occurring at constrictions. The largest fall in the water surface was probably at the Swan Falls constriction, where the modeled profile drops about 25 m. Critical or supercritical flow conditions were calculated for the Swan Falls constriction and at the constriction immediately upstream of Walters bar (Fig. 39).

Jarrett and Malde (1987) used a step-backwater computer program similar to HEC-2 to obtain an estimate of 0.793 to 1.02 $\times 10^6 m^3 \cdot sec^{-1}$ for peak discharge through the Swan Falls constriction. Considering several factors, including possible widening of the constriction during the flood, they proposed that a discharge of $0.935 \times 10^6 m^3 \cdot sec^{-1}$ is a most likely estimate for peak discharge. For model runs with hydraulic parameters similar to those used in this study, they estimated a discharge of $0.99 \times 10^6 m^3 \cdot sec^{-1}$, 15 to 20% greater than the 0.82 to $0.85 \times 10^6 m^3 \cdot sec^{-1}$ estimate of peak discharge obtained in this study. Part of the discrepancy may be the result of using different sets of cross sections to characterize the channel geometry. However, it more likely relates to the strategy of this study, which was to define a *minimum* estimate of peak discharge. For the Swan Falls reach, where flow had high velocities near the important high-water evidence, I have compared the stage evidence to the energy-surface profile, instead of the calculated water-surface profile because of possible runup effects (Fig. 39).

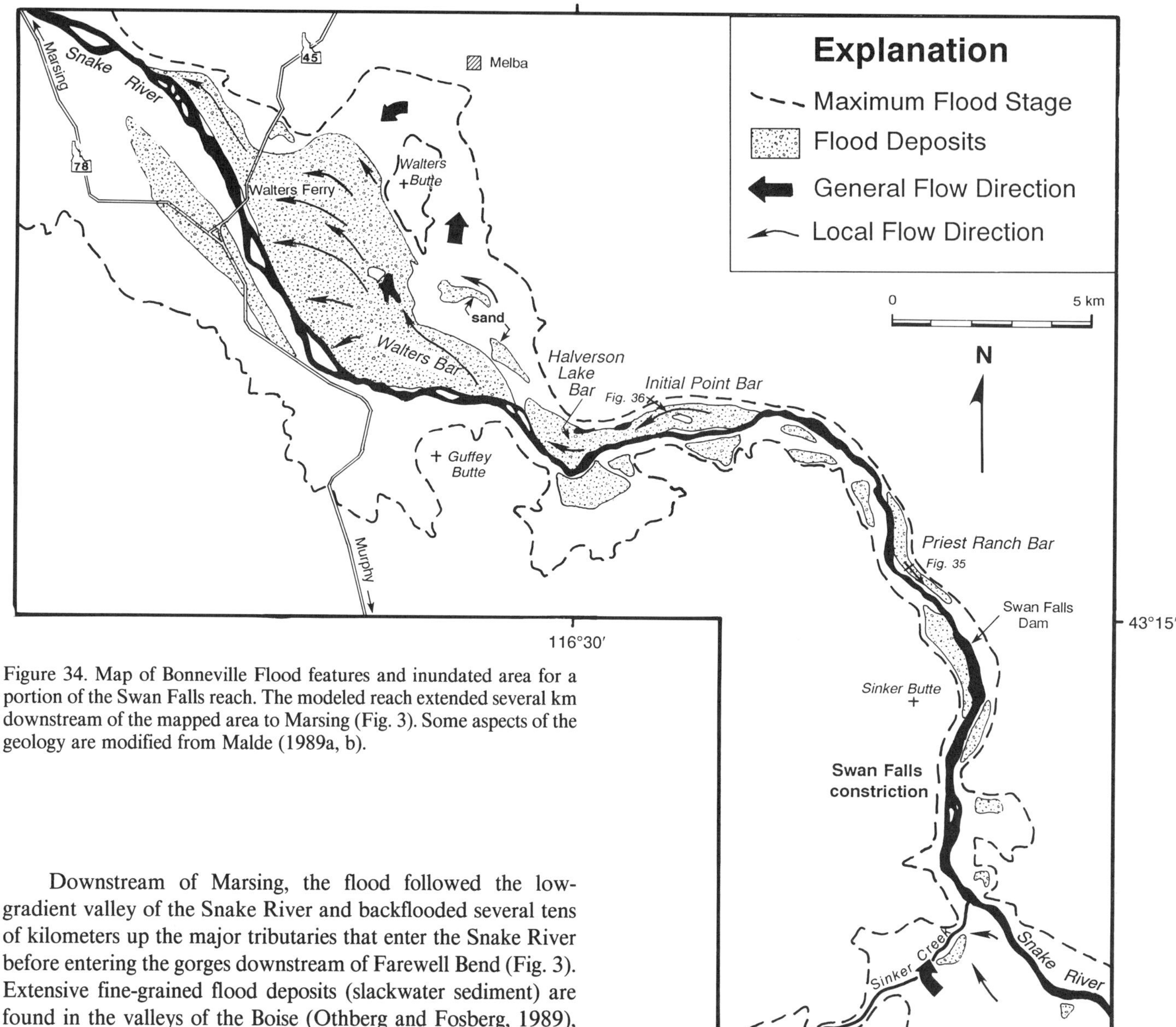

Figure 34. Map of Bonneville Flood features and inundated area for a portion of the Swan Falls reach. The modeled reach extended several km downstream of the mapped area to Marsing (Fig. 3). Some aspects of the geology are modified from Malde (1989a, b).

Downstream of Marsing, the flood followed the low-gradient valley of the Snake River and backflooded several tens of kilometers up the major tributaries that enter the Snake River before entering the gorges downstream of Farewell Bend (Fig. 3). Extensive fine-grained flood deposits (slackwater sediment) are found in the valleys of the Boise (Othberg and Fosberg, 1989), Payette, and Owyhee Rivers. The flood profile dropped very little through this reach; water was apparently ponded to an altitude of at least 746 m (2,447 ft) behind the constriction near Farewell Bend.

Farewell Bend. The constriction at Farewell Bend is where the Snake River enters the series of gorges that the river follows along the Oregon-Idaho and Washington-Idaho borders (Fig. 40). Downstream of Farewell Bend, preserved flood deposits are sparse and maximum stage evidence is difficult to discern confidently. This is partly due to the narrowness of the flood route but is primarily the result of the extremely steep and active slopes along the margins of the gorge. Calculating water-surface profiles for the Farewell Bend reach was done for the sole purpose of obtaining a discharge estimate associated with erosional features upstream of the constriction at Farewell Bend that indicate a minimum ponding level of 746 m (2,447 ft) at peak discharge.

Flow was modeled through 68 cross sections over a distance of 50 km. A water-surface profile associated with a discharge of 0.65×10^6 $m^3 \cdot sec^{-1}$ results in a ponding level behind the constriction consistent with the altitudes of the highest flood features (Fig. 41). A discharge of 0.82×10^6 $m^3 \cdot sec^{-1}$ results in a predicted ponding level of 760 m (2,494 ft), the maximum stage at the downstream end of the Swan Falls reach, and probably provides a reliable upper limit for an estimate of the discharge that entered the Farewell Bend reach.

Downstream of the Farewell Bend reach, the Snake River continues a deeply incised course within lava flows of the Weiser embayment of the Columbia River Basalt Group until it enters

Figure 35. Large and rounded boulders deposited by the Bonneville Flood at the Priest Ranch bar near Swan Falls (photograph location shown in Fig. 34). Note person for scale; many of the boulders have long diameters that exceed 10 m. View is southeast (upstream) toward Swan Falls Dam that is faintly visible in the background.

Figure 36. View southeast of the Initial Point bar (photograph location shown in Fig. 34). The photograph was taken from the canyon rim at an altitude near the maximum flood stage. A large eddy bar (Fig. 37) is visible across the river from the outcrop of eroded basalt. USGS photograph by Harold E. Malde (station 78-M-93).

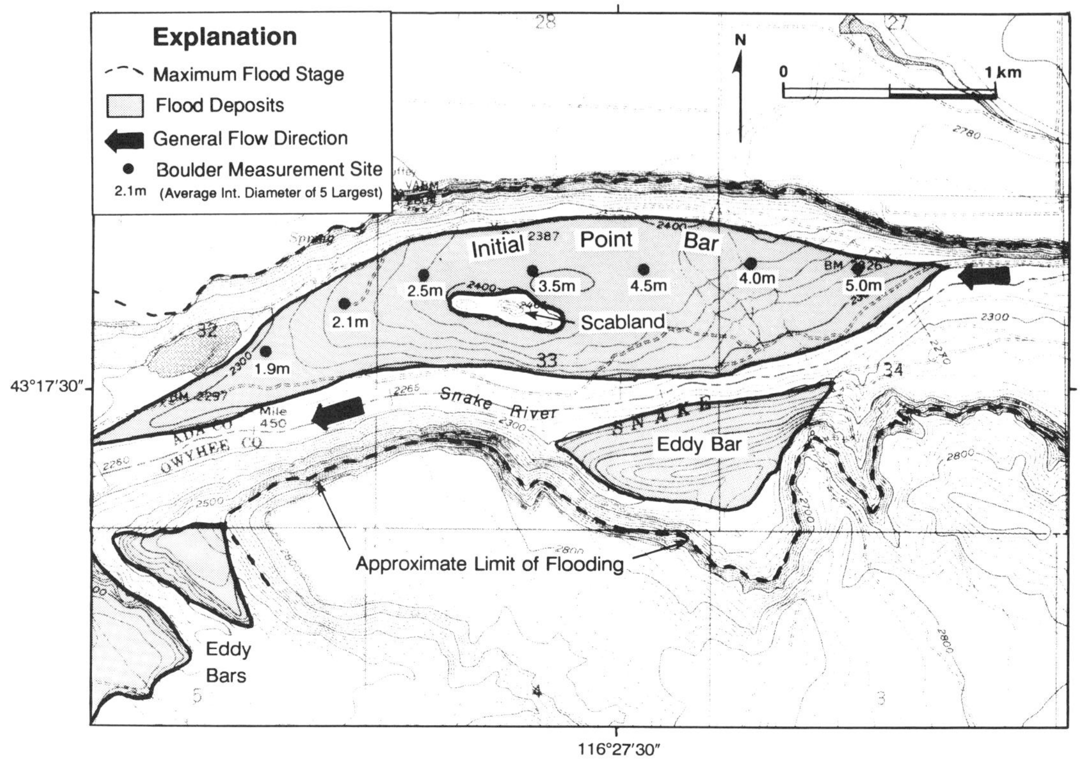

Figure 37. Topography and flood features near the Initial Point bar (Fig. 34). The Initial Point bar is a 4-km-long longitudinal bar deposited in a locally wider portion of the canyon. Large eddy bars formed in canyon reentrants south of the Snake River. Topography from the USGS (ID) Initial Point 7.5-min Quadrangle.

the Permian-Triassic Seven Devils Complex of volcanic and volcaniclastic rocks in Hells Canyon. Bonneville Flood deposits are only locally preserved or exposed in some of the wider portions of the canyon or where local placer and gravel operations have removed post-flood colluvium. Extensive flood deposits are not found downstream until the middle of Hells Canyon where large bars have been deposited in wider reaches of the canyon.

Pittsburg Landing and Dug Bar reaches (Hells Canyon). Two separately modeled, but contiguous, reaches were analyzed in the central part of Hells Canyon (Fig. 42). Hells Canyon is a steep and narrow V-shaped gorge with the Snake River generally occupying the entire 50- to 150-m-wide canyon bottom (Fig. 9). The river consists of a series of straight, fault-controlled, segments separated by abrupt bends. Major accumulations of flood deposits along the canyon bottom are restricted to wider locations; some of the larger bar complexes are at Johnson Bar, Big Bar, Pittsburg Landing (Fig. 43), Roland Bar, and Dug Bar (Fig. 44). In some segments of the canyon, there are deposits of angular and poorly sorted gravel that were left at the flow margins, and where material in suspension was deposited in back-flooded tributary valleys (Figs. 43, 44).

The altitudes of the suspended-load deposits were used to determine minimum constraints on the flood profile. At some locations the tops of these deposits were surveyed; however, in most locations, they were mapped from aerial photos and their altitudes determined from topographic maps. Consequently, the water-surface profile of the flood through Hells Canyon is only locally determined with precision. Flood gravel near Johnson Bar indicates a maximum stage of at least 541 m (1,775 ft) and a flow depth of 175 m (Fig. 45). Downstream at Pittsburg Landing, flood deposits indicate that the maximum stage exceeded 507 m (1,665 ft) (Fig. 46). Farther downstream, in the Dug Bar reach, the highest gravel deposits are at an altitude of 396 m (1,300 ft) near the confluence with the Salmon River.

In combination, step-backwater calculations were made for 180 cross sections over a distance of 70 km for the two reaches (Figs. 45, 46). For the Pittsburg Landing reach (Fig. 45), a water-surface profile associated with a discharge of 0.57×10^6 $m^3 \cdot sec^{-1}$ gives a good match to the geologic evidence of maximum stages. In calculating this profile, critical flow was assumed at the Pittsburg Landing constriction at the downstream end of the reach. In contrast, for the Dug Bar reach, calculated profiles with the same discharge and reasonable hydraulic parameters give water-surface profiles that are substantially higher than the geologic evidence of maximum flow stages (Fig. 46). This result is similar to those obtained from other steep reaches along the flood

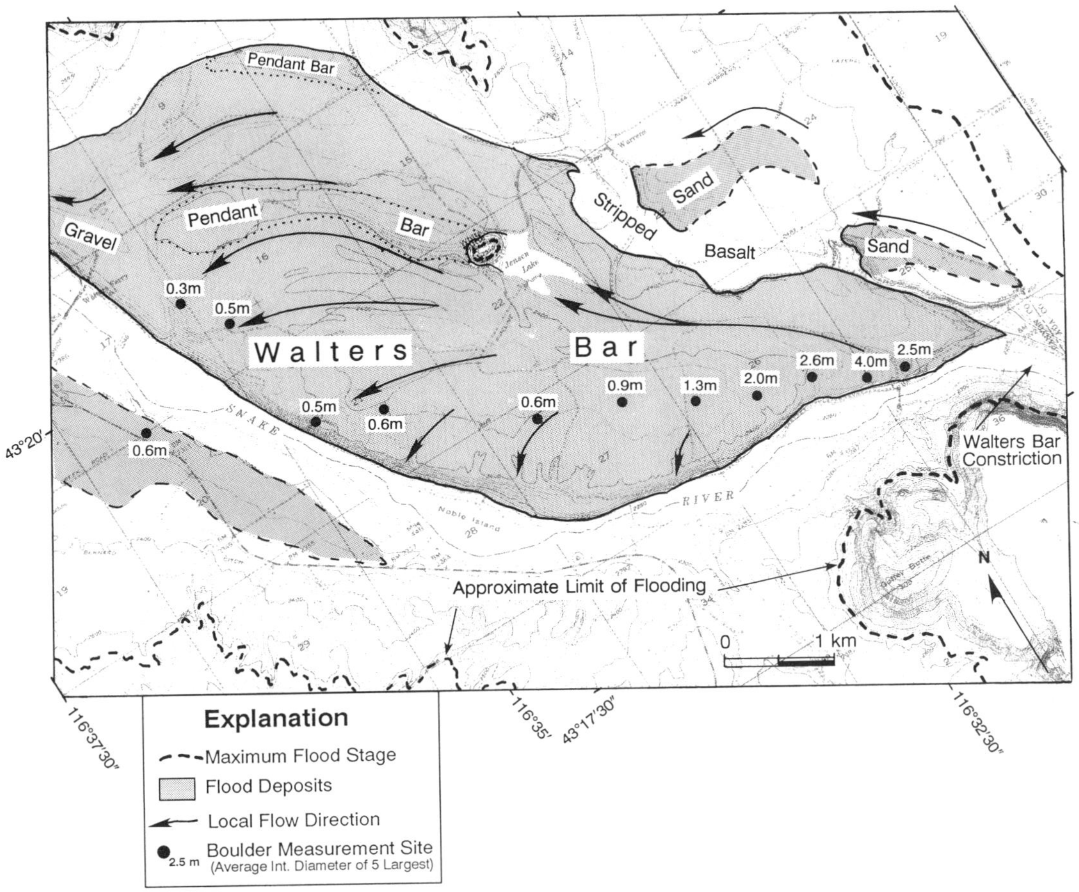

Figure 38. Flood features, topography, and boulder measurement locations at Walters bar within the Swan Falls reach (Fig. 34). The Walters bar is a large expansion bar formed where the flow exited the confined canyon downstream of Swan Falls and entered a wider, ponded reach downstream. The boulder measurement site of smaller clasts at the apex of the Walters bar is apparently on an inset surface formed later in the flow, after deposition of the main body of Walters bar. Local flow directions are defined by channels incised into the bar surface. Some aspects of the geology from Malde (1989a). Topographic base from the Walters Butte (ID) USGS 7.5-min Quadrangle.

where, apparently, the step-backwater method applied with standard choices for energy-loss coefficients does not adequately account for all of the energy lost by the flow.

The downstream end of the modeled Hells Canyon reaches is near the confluence of the Snake and Salmon Rivers (Fig. 3). Farther downstream, the Snake River remains confined within a deep and narrow gorge until near the confluence with the Grande Ronde River, where the canyon widens and the gradient declines as the Snake River exits the resistant rocks of the Seven Devils complex and enters the valley incised into lava flows of the Clearwater embayment of the Columbia River Basalt Group.

China Garden to Lewiston. The Lewiston reach is a 55-km segment of the flood route between China Garden and Lewiston (Fig. 47). This reach encompasses the transition between Hells Canyon and a wider valley downstream. There are abundant deposits along this reach, especially at the downstream end near Lewiston where there are large point-bar complexes (Fig. 48) (Webster and others, 1982; Kehew, 1977). High-level eddy deposits and erosional features serve to constrain the maximum stage of the Bonneville Flood at several locations along the reach (Fig. 49).

The evidence for maximum stages and the flow reconstructions indicate ponding in the Lewiston Basin, probably resulting from downstream constrictions. Maximum altitudes of flood features are consistently at altitudes of 317 to 320 m (1,040 to 1,050 ft) for several kilometers upstream of the Snake-Clearwater confluence (Fig. 49). At the upstream end of the study reach, at China Garden, gravel deposited by the Bonneville Flood reaches an altitude of 396 to 408 m (1,300 to 1,340 ft).

Step-backwater profiles were calculated through 80 cross sections over a distance of 55 km. A minimum discharge of 0.57 to 0.62×10^6 $m^3 \cdot sec^{-1}$ is consistent with the geologic evidence of maximum stages (Fig. 49). This estimate is primarily constrained by the altitude of the deposit at China Garden. Ten

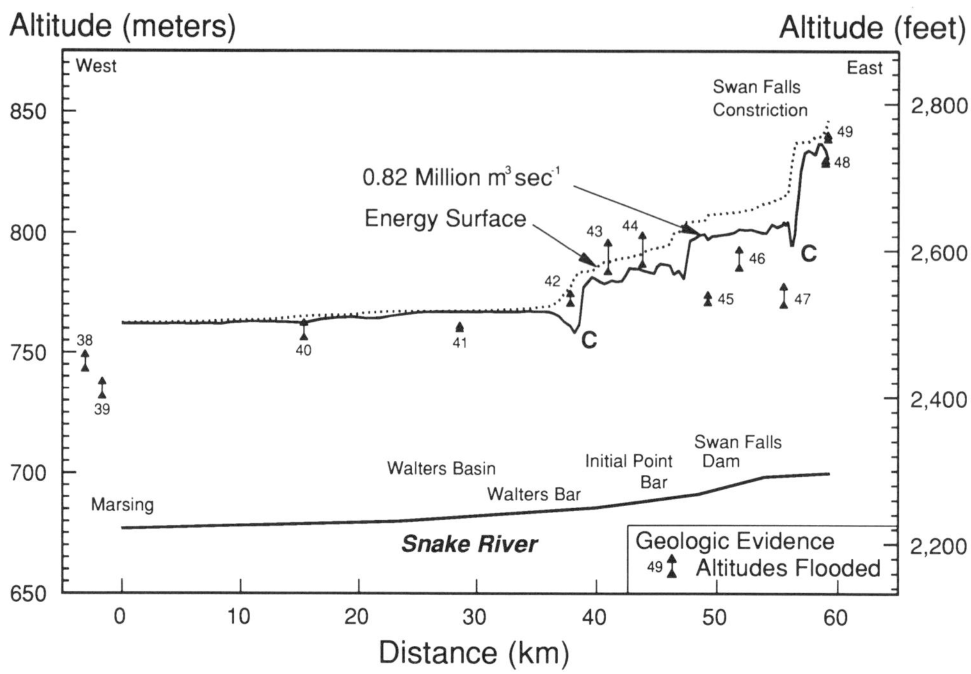

Figure 39. Geologic evidence for maximum flood stages and calculated water-surface and energy profiles for the Swan Falls reach (Highwater evidence #38 and #39 are downstream and provide constraints on the minimum flood height at the downstream end of the reach). A discharge of 0.82 million $m^3 \cdot sec^{-1}$ is the minimum that could have emplaced the high water evidence.

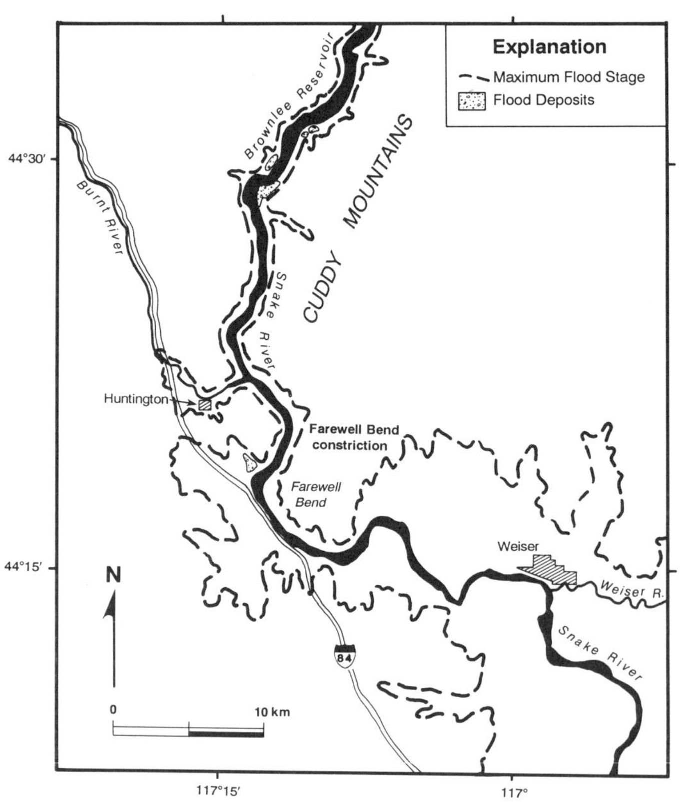

Figure 40. Major flood deposits and the inundated area near Farewell Bend.

kilometers downstream from China Garden, critical flow conditions were calculated for the Grande Ronde constriction. Farther downstream, the gradient of the calculated water-surface profiles diminish in a manner consistent with the geologic evidence of maximum stages.

Discussion of modeling results

The step-backwater flow reconstructions provides precise minimum estimates of maximum flood discharge for several points along the flood route. The most reliable discharge estimates were obtained where there are good constraints on maximum stage altitudes upstream of major constrictions. In general, the discharge estimates are mutually consistent and provide a basis for evaluating the flood hydrology. In reaches where the modeled profiles match the water surface defined by the high-water indicators, the step-backwater calculations also provide a quantitative basis for analyzing the local hydraulic conditions.

Quality of the modeling. The similarity between the modeling results and the geologic evidence of maximum stage provides an indication of how well the step-backwater calculations describe the actual flow. The best matches were achieved for the Swan Falls, Hagerman, and the Pocatello–Red Rock Pass reaches. These reaches are all similar in that they had long sections of wide, hydraulically ponded flow separated by short, steep reaches of unimpeded critical to supercritical conditions. The flow was energetic enough in these reaches to leave abundant

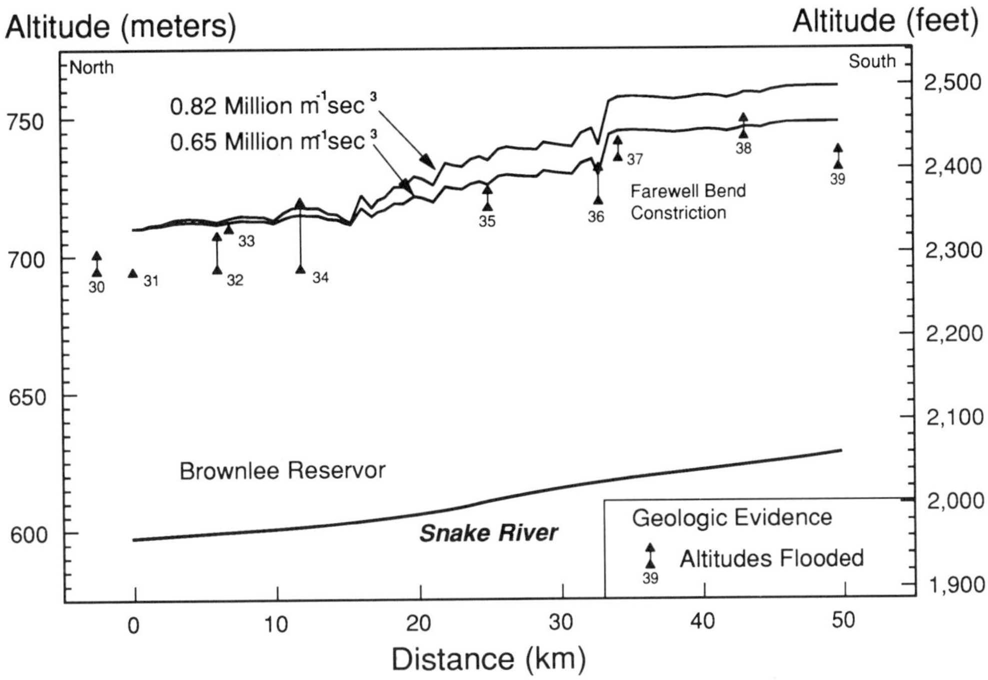

Figure 41. Geologic evidence of maximum stages and calculated profiles for the Farewell Bend reach. A discharge of 0.65 million $m^3 \cdot sec^{-1}$ results in a calculated water-surface profile that matches or exceeds all of the high water evidence. A discharge of 0.82 million $m^3 \cdot sec^{-1}$ results in a calculated water-surface profile consistent with the evidence of the maximum stage at the downstream end of the Swan Falls reach.

evidence of maximum stages. Moreover, there are sufficient constricted segments where critical flow was achieved that the reconstructed water-surface profiles are uniquely defined with respect to discharge and can be precisely related to the flood evidence. This finding is consistent with Jarrett's (1987, p. 66) observation that "the critical-depth method of computing peak discharge provides the most reasonable peak discharges in higher-gradient streams."

Reaches in other hydraulic environments were not as suitable for paleohydraulic analysis. Analysis of the reach between Raft River and the Burley Basin illustrates the problems in modeling wide, low-gradient reaches. The step-backwater calculations may be accurate; however, it is difficult to assign precise discharge estimates on the basis of geologic evidence because of the paucity of high-level deposits. There were not enough erosional and depositional features left by the slow-moving flow to form a geologically well-defined water-surface profile with which to evaluate the calculated profiles. Furthermore, step-backwater flow calculations in channels without sections of critical flow are more sensitive to uncertainties in energy-loss coefficients because of the potentially large cumulative effect resulting from systematically under- or overestimating these coefficients.

More perplexing results were those from the very steep reaches such as Hells Canyon (especially the Dug Bar reach) and the Murtaugh reach. These narrow, bedrock-bounded channels would seem to be ideal locations for step-backwater modeling. However, the calculated water-surface profiles associated with appropriate discharges are substantially higher than the geologic evidence of maximum stages (Figs. 24, 46). Some of the discrepancy between the geologic evidence and the modeled profiles in Hells Canyon could be attributed to the scarcity of high-level flood features. However, the Murtaugh reach has abundant highwater evidence that apparently tightly bracket the flood profile. More likely is that the step-backwater calculations are providing unreliable results in these locations. The assumptions of gradually varied flow may not be tenable in these narrow and steep channels, and there may be much more energy loss in portions of these reaches than is accounted for in the flow reconstructions.

Moreover, long sections of flow may have experienced near-critical flow conditions. The well-constrained Murtaugh profile suggests that long segments of the flow were near critical depth (Fig. 24). The downstream portions of the Eden Channel and the Dug Bar reach also have evidence of maximum stages at altitudes near critical depth for an appropriate discharge (Figs. 25, 46). Apparently, flow energy was expended in such a manner, by mechanisms not accounted for in the step-backwater calculations, that conditions close to minimum specific energy (critical depth) were maintained. This is reasonable because local energy losses associated with hydraulic jumps at flow transitions (between su-

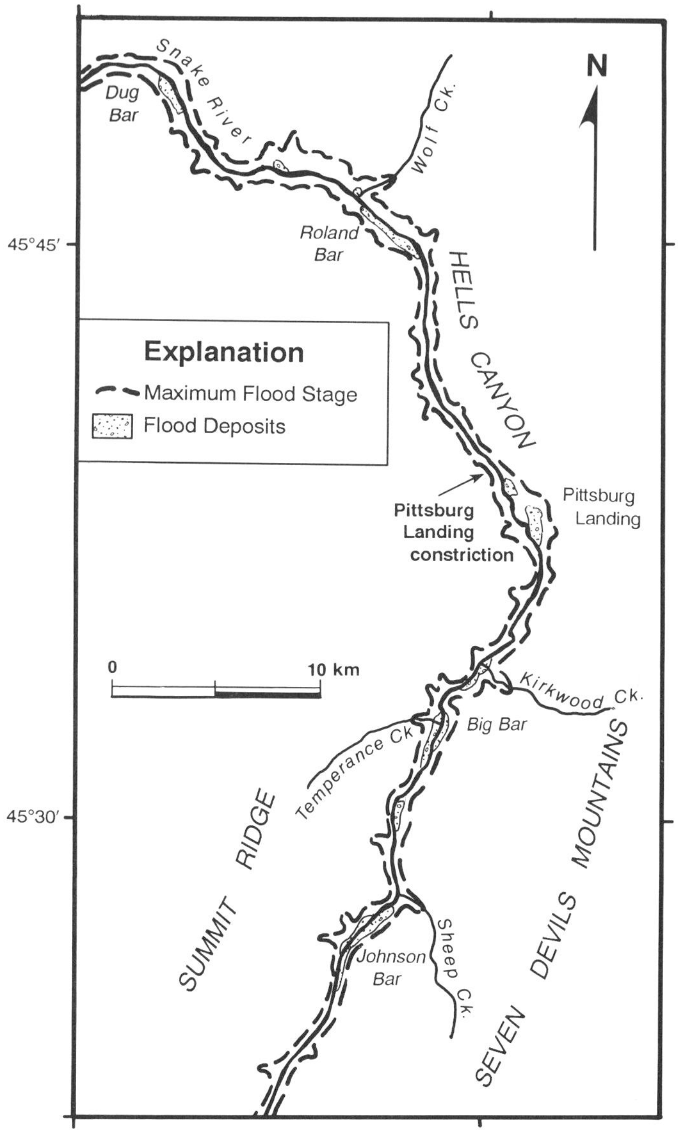

Figure 42. Map of major flood features and the approximate area of inundation within the central portion of Hells Canyon.

percritical and subcritical flow) can be large and are not accounted for by the step-backwater calculations. Moreover, energy dissipation by this process would act efficiently to maintain smaller flow depths in steep and narrow channels with irregular boundaries, and would counteract the tendency toward the large, hydraulically ponded flow depths predicted by the step-backwater calculations.

Bonneville Flood hydrology. Minimum discharge values were estimated for all of the reaches except for the Dug Bar reach despite local problems in matching the calculated water-surface profiles with those defined by the geologic evidence (Table 1, Fig. 3). These discharge values are mutually consistent: peak discharge decreases downstream from a maximum value of about 1.0×10^6 $m^3 \cdot sec^{-1}$ at the outlet near Red Rock Pass to 0.57 to 0.62×10^6 $m^3 \cdot sec^{-1}$ near Lewiston at the downstream end of the studied flood route. For the Eden channel and Murtaugh reaches, where the flow diverged along two paths, the combined discharge estimate is consistent with upstream and downstream estimates where all the flow was within a single channel. For comparison, the largest historic discharge on the Snake River was about 11,300 $m^3 \cdot sec^{-1}$ near Lewiston in 1884. Upstream, near the Oxbow, the maximum gaged discharge has been 2,266 $m^3 \cdot sec^{-1}$.

The total volume of the flood was the prism of water between the Provo and Bonneville levels of Lake Bonneville (Fig. 4). Hence, a total flood volume of 4,750 km^3 can be calculated from the hypsometry of the Bonneville Basin (Fig. 50) (Jarrett and Malde, 1987). This volume is more than twice the 2,200 km^3 of water that emptied from Glacial Lake Missoula (from the stage-volume relationship for Glacial Lake Missoula provided by Craig [1987]) (Fig. 1) in the largest of the Missoula Floods. Despite a much larger total volume, the Bonneville Flood had a maximum discharge an order of magnitude less than the largest Missoula Flood of about 17×10^6 $m^3 \cdot sec^{-1}$ (O'Connor and Baker, 1992).

Some aspects of the breach hydrograph and downstream migration of the flood wave can be evaluated by hydraulic flow routing. Hydraulic flow routing requires numerical solution of the dynamic wave (St. Venant) equations for unsteady flow. Solutions of these equations can be used to predict the downstream propagation of a flood wave, given a specified hydrograph at the upstream end of the channel and the channel geometry. Bedient and Huber (1988, p. 209–277) have provided a summary of the physical and mathematical basis of several flood-routing techniques. The routing model used for the Bonneville Flood was the National Weather Service DAMBRK Model (Fread, 1988), a computer model designed to simulate breach hydrographs from dam failures and to predict the behavior of those hydrographs as they move downstream. Fread (1988) described the solution procedure employed in DAMBRK and the advantages of this particular algorithm for predicting hydrographs resulting from dam failures.

Only the portion of the model that predicts the downstream movement of the floodwave was applied to the analysis of the Bonneville Flood. Forty-five cross sections were measured from topographic maps to describe the flood route between Red Rock Pass and Farewell Bend (Fig. 3). These cross sections were differ-

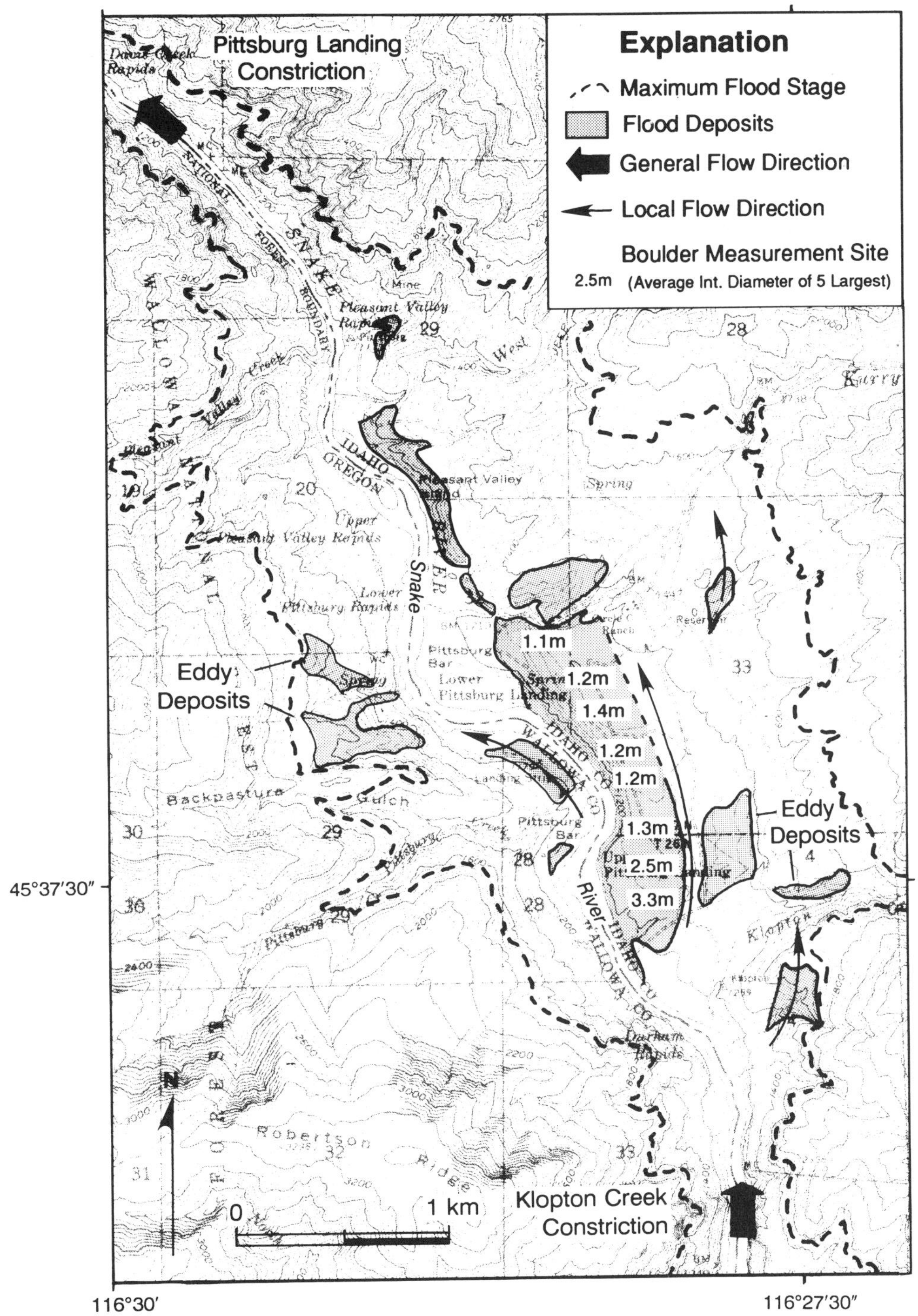

Figure 43. Topography, flood features, and the approximate area of inundation in the Pittsburg Landing area (Fig. 42). Deposition of the large bars was probably the result of local hydraulic ponding behind the Pittsburg Landing constriction. Topography from the Grave Point and Kirkwood Creek (ID-OR) 7.5-min USGS Quadrangles. Geology mapped in conjunction with Tracy Vallier of the US Geological Survey.

ent than those describing the channel geometry for the step-backwater modeling: whereas the cross sections describing the channel for the step-backwater calculations were specifically placed and oriented to describe the portions of the channel that conveyed flow, cross sections used for the routing also need to describe areas of flow storage and backwater areas. Because of numerical limitations, the number of cross sections was limited and only a very general portrayal of the channel system could be satisfactorily modeled. Consequently, the results of the hydraulic modeling can be interpreted only in a general manner.

For the Bonneville Flood, the step-backwater calculations provide discharge estimates at several locations downstream from the breach at Red Rock Pass. In addition, the total volume of the flood is known from the hypsometry of the Lake Bonneville

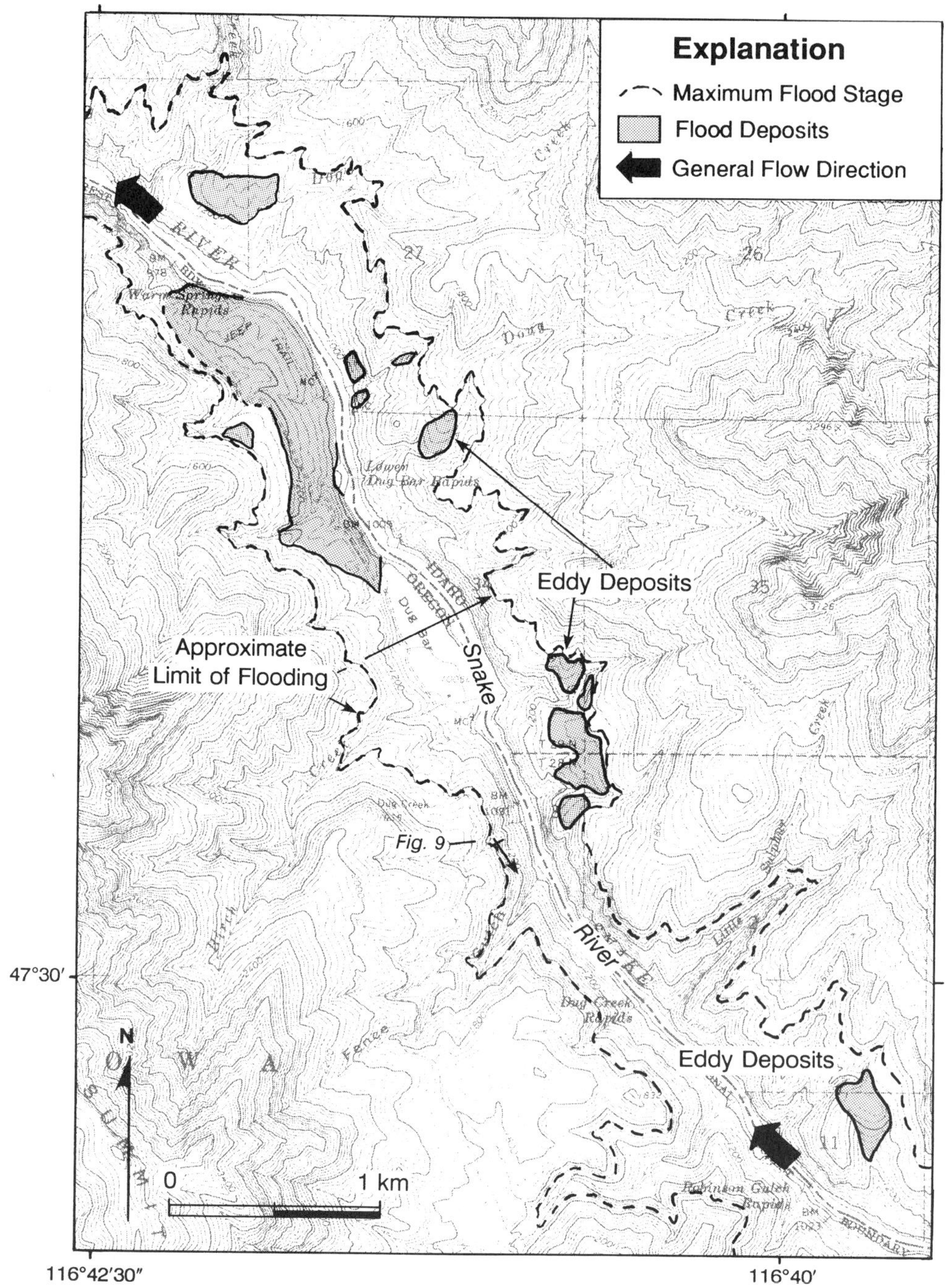

Figure 44. Topography, flood features, and the approximate limits of inundation near Dug Bar (Fig. 42). Several gravel deposits, probably derived from suspended load, were deposited in canyon reentrants and tributary valleys. Topography from the Cactus Mountain (ID-OR) 7.5-min USGS Quadrangle.

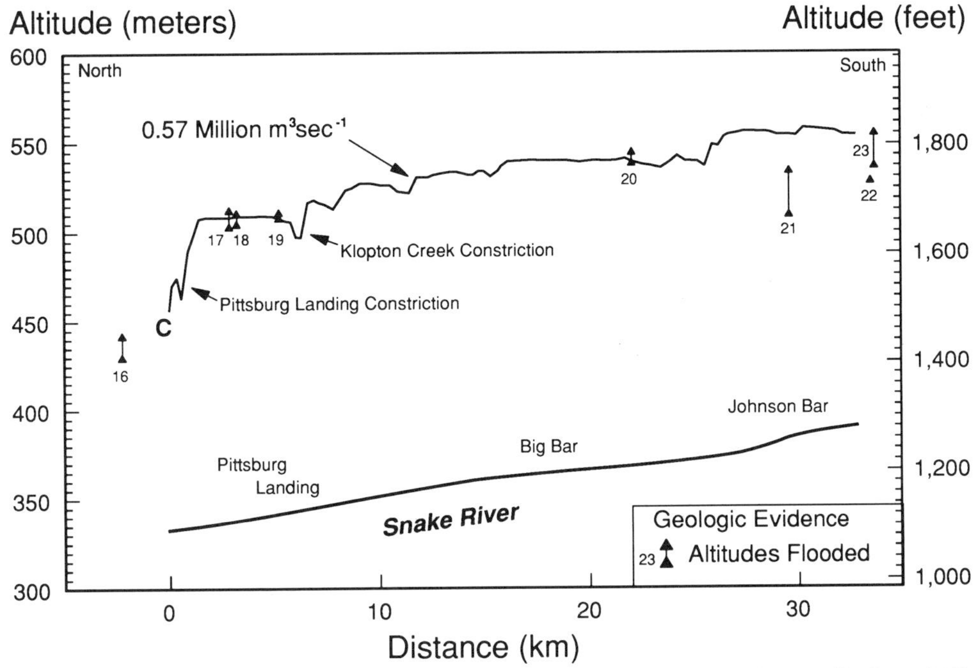

Figure 45. Geologic evidence of maximum stages and a calculated water-surface profile for a 0.57-million $m^3 \cdot sec^{-1}$ discharge within the Pittsburg Landing reach in Hells Canyon. This reach is bounded by the Pittsburg Landing constriction to the north and Johnson Bar to the south (Fig. 42). Discharges of 0.57 to 0.62 million $m^3 \cdot sec^{-1}$ result in calculated profiles that just exceed the altitudes of the highest gravel deposits along the reach. Critical flow was assumed at the Pittsburg Landing constriction.

Basin. Not known are the shapes of the hydrographs at the study reaches and the downstream rate of migration of the flood peak. To estimate these properties, several hydrograph shapes were proposed, consistent with the known flood volume of 4,750 km^3 and the 1.0×10^6 $m^3 \cdot sec^{-1}$ estimate of peak discharge at Red Rock Pass, and routed down the flood path to Farewell Bend. Predicted peak discharges at downstream locations were compared with the discharges estimated from the step-backwater analyses.

Figure 51 shows the results of a routing trial that gave downstream peak discharges consistent with the discharge estimates from the step-backwater calculations. These hydrographs should only be viewed as being representative of possible conditions; other combinations of postulated hydrographs and assigned roughness coefficients also give results consistent with the discharge estimates. Nevertheless, all trials that successfully matched the step-backwater discharge estimates downstream required that the hydrograph at Red Rock Pass have a sharp peak early in the flow. The postulated hydrograph for Red Rock Pass portrayed in Figure 51 peaked 100 hr (4 days) after the beginning of substantial flow, that is, after downcutting began at the Marsh Creek fan. For this hydrograph, the total flow duration at Red Rock Pass was about 7,200 hr (300 days), with flow exceeding 0.5×10^6 $m^3 \cdot sec^{-1}$ for about 430 hr (18 days). At Swan Falls, the peak discharge is simulated as attenuating to 0.82×10^6 $m^3 \cdot sec^{-1}$, matching the estimate of the step-backwater modeling, about 5 days after the peak at Red Rock Pass. The routing results for this trial indicate that, at Swan Falls, a discharge within 10% of the peak discharge was sustained for about 150 hr (6 days), lending credence to the steady-flow assumption intrinsic to the step-backwater method.

The peak discharge at Farewell Bend resulting from the flow routing portrayed in Figure 51 was 0.78×10^6 $m^3 \cdot sec^{-1}$; within the bracketing estimates from the step-backwater calculations (Table 1), but 26% higher than the best estimate of 0.62×10^6 $m^3 \cdot sec^{-1}$. Even greater attenuation than shown in Figure 51, especially between Swan Falls and Farewell Bend, would probably have been simulated if the peak of the input hydrograph at Red Rock Pass was placed earlier in the hydrograph; however, numerical instability caused by simulating more rapidly rising hydrographs prevented successful routings of input hydrographs with peaks earlier in the flow than that of Figure 51. The simulated floodwave peaked at Farewell Bend 320 hr (13 days) after the peak at Red Rock Pass, indicating that the flood peak migrated downstream at about 1 $m \cdot sec^{-1}$, a value that seems low considering the much higher velocities calculated from the step-backwater reconstructions.

A flood hydrograph with a peak early in the flow is particularly susceptible to attenuation because the flow peak arrives prior to filling of areas of local channel and backwater storage. In

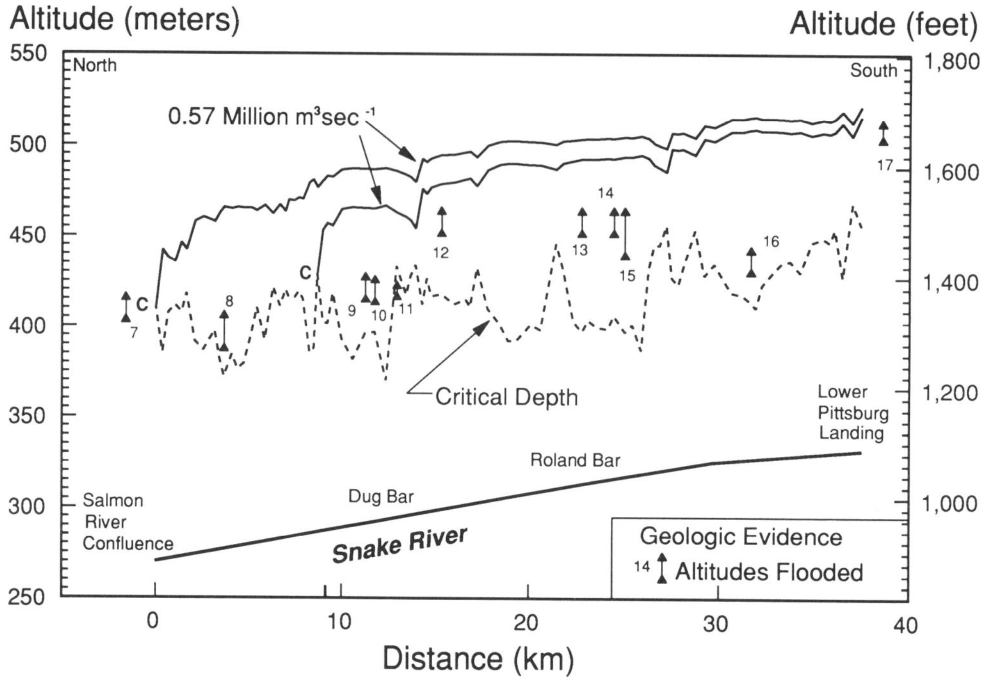

Figure 46. Calculated water-surface profiles, critical depth profile, and geologic evidence for maximum stages in the Dug Bar reach within Hells Canyon. This reach encompasses the Snake River canyon from the Salmon River confluence upstream to the Pittsburg Landing constriction (Fig. 3). Both water-surface profiles are those associated with a 0.57-million $m^3 \cdot sec^{-1}$ discharge. The lower profile is calculated on the basis of assumed critical flow conditions at a cross section immediately downstream of Dug Bar. The upper profile was started from critical depth at a constriction near the confluence with the Salmon River. Both of the water-surface profiles are substantially higher than the geologic evidence of maximum stages, especially near Dug Bar and Roland Bar where there are abundant high-level deposits with well-determined altitudes. Nevertheless, they do converge to an appropriate altitude near Lower Pittsburg Landing. Similar to the Murtaugh reach, the geologic evidence indicates that there were long reaches of near critical flow, particularly between Dug Bar and the Salmon River confluence.

such situations, a greater portion of the discharge will be delayed as local backwater areas are flooded. Consistent with the results of the hydraulic modeling and the step-backwater estimates of discharge, this was probably an important factor in the Bonneville Flood. It is most likely that discharge peaked early in the flood—after full incision of Red Rock Pass, but while the lake level behind the outlet was still high. The reach between Swan Falls and Farewell Bend was probably especially effective in attenuating the flow because of the vast storage associated with the wide alluvial valley of the Snake River and backflooding up several of the major tributaries (Fig. 3).

Flow energy expenditure. The release of 4,750 km^3 of water at an altitude of 1,552 m and its conveyance to sea level resulted in a tremendous release of energy. It is this energy that performed the large amount of geomorphic work done along certain segments of the flood route. The total energy expended by the flood in going from an altitude of 1,552 m to sea level was 7.2×10^{19} Joules, about equivalent to the annual US energy production. If all this energy was converted to heat, the temperature of the entire flood volume would have been raised by 3.6°C in descending to sea level.

Only a small portion of the total flow energy would have been applied to geomorphic work. Most of it would have been converted to heat through viscous dissipation and boundary friction. Some energy was also undoubtedly lost from the flow as seismic and acoustic energy. Nevertheless, flow energy expenditure (and the rate of energy expenditure, power) provides a useful index of the ability of the flood to perform geomorphic work.

Both the high-water indicators and the flow reconstructions show the remarkable spatial variation in energy expenditure over the flood route. For the scale of the studied flood route and within the individual study reaches, the local mechanical energy of the flow at peak discharge is approximately represented by the altitude of the water surface. The high-water indicators document that the water-surface profile (and the associated energy surface) was stepped; reaches of intense energy loss were separated by long reaches of relatively tranquil flow with little energy loss (Fig. 10).

For the modeled reaches, the variation in the rate of energy expenditure can be demonstrated from the step-backwater calculations for each modeled flow increment. Energy loss between cross sections is represented by h_e, the head loss. This is calculated

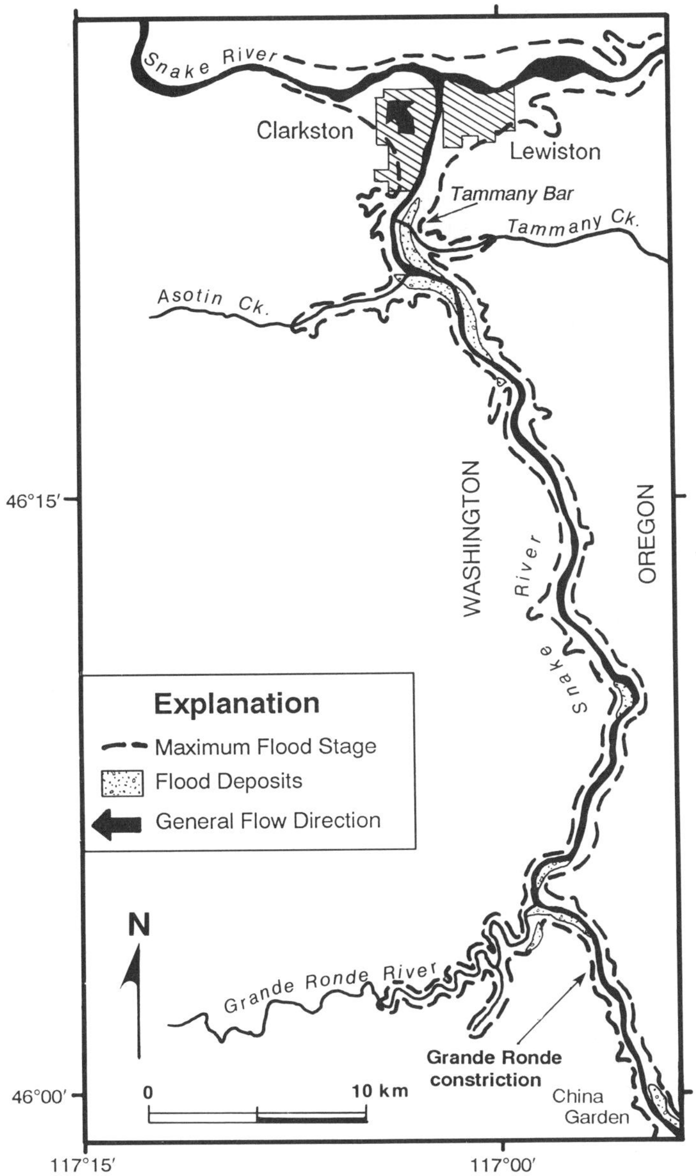

Figure 47. Map of major flood features and the approximate area of inundation within the Lewiston reach.

with equation (2) between each pair of modeled cross sections (Fig. 12), and represents the total estimated decrease in mechanical energy within the channel increment. Considering all of the modeled cross-section pairs, which encompass 50% of the distance and 75% of the energy loss for the flood route between Red Rock Pass and Lewiston, there is substantial nonuniformity in energy dissipation (Fig. 52). Sixty percent of the flow energy was expended within 10% of the modeled distance.

Some aspects of the flood hydraulics. The step-backwater reconstruction of flow conditions allows for analysis of the spatial variations in some of the key flow properties. Some of the geomorphically important variables are flow depth (d), velocity (v), energy slope (S_e), boundary shear stress (τ), and unit stream power (ω). Flow depth, velocity, and energy slope values are obtained directly from the modeling results at each cross section and can be used to derive boundary shear stress and unit stream power magnitudes.

Local values of these variables *within* a cross section can also be estimated from the step-backwater calculations (Fig. 53). Local flow depth is determined by subtracting the local ground altitude from the calculated water-surface altitude. A highly generalized velocity distribution across each cross section is calculated on the basis of local flow depth, local estimated roughness (Mannings n), and energy slope. These velocity values are vertically averaged estimates for the flow column. Furthermore, they are calculated on the basis of the average energy slope for the portion of the cross section conveying downstream-directed flow; if the energy slope varies across the cross section, local velocity estimates could be in error. Consequently, these velocity values must be viewed as only an approximation of actual local flow conditions.

Boundary shear stress is the tangential force exerted by the flow, per unit area, on the flow boundary. It is a parameter commonly used to evaluate the ability of a flow to entrain and transport sediment (e.g., Rubey, 1938; Baker, 1973; Williams, 1983; Costa, 1983; Baker and Costa, 1987; Komar and Carling, 1991). Local values of boundary shear stress can be calculated by balancing fluid forces experienced by the bed:

$$\tau = \gamma d S_e \qquad (4)$$

where γ is the specific weight of the fluid, here taken to be 9,800 $N \cdot m^{-3}$ (clear water).

Bagnold (1966, 1977, 1980) suggested that unit stream power is strongly related to a flow's ability to transport sediment and to perform geomorphic work. The term "unit stream power" is used here as defined by Bagnold (1966, p. 5–6), and refers to the power (time rate of energy expenditure) per unit area of bed. Later workers have also noted strong relations between local stream power magnitude and flow competence (Williams, 1983; Costa, 1983; Baker and Costa, 1987). One expression of stream power (ω) is:

$$\omega = \tau v. \qquad (5)$$

An important aspect of the derivation of shear-stress and unit stream power values is their direct dependence on the local energy slope (S_e). However, workers using these parameters for analysis of sediment transport have frequently made the simplifying assumption that the channel slope or the water-surface slope is an adequate proxy for the energy slope. This can introduce large errors for conditions of nonuniform flow; Costa (1987) reported substantial differences between channel slopes and energy slopes for flash floods in small watersheds. The flow reconstructions for the Bonneville Flood also indicate large local discrepancies (e.g., Figs. 30, 39). A key property of the step-

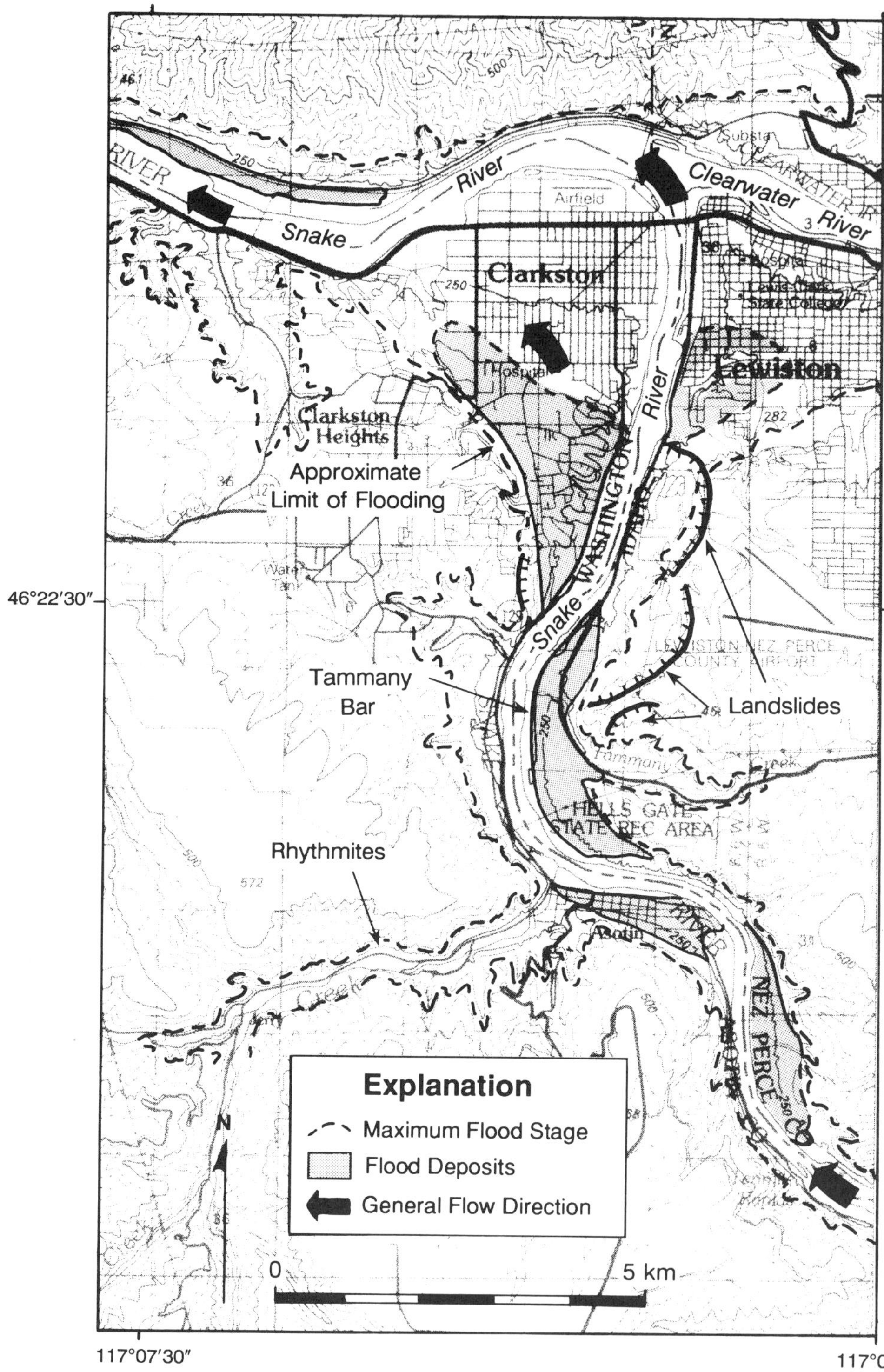

Figure 48. Map of flood features near Lewiston, Idaho (Fig. 47). Extensive deposition of Bonneville Flood sediment near Lewiston resulted from hydraulic ponding behind downstream constrictions. Graded beds of sand and silt (rhythmites), probably deposited by the Missoula Flood(s), cap Bonneville Flood deposits at Tammany bar (Fig. 74) and are found in some of the tributary valleys. The large landslides flanking the Snake River have not been active historically (Kehew, 1977) and may have been caused by passage of the Bonneville Flood. Some elements of the geology are taken from Webster and others (1982). Metric unit topography from the USGS Clarkston Quadrangle scale 1:100,000.

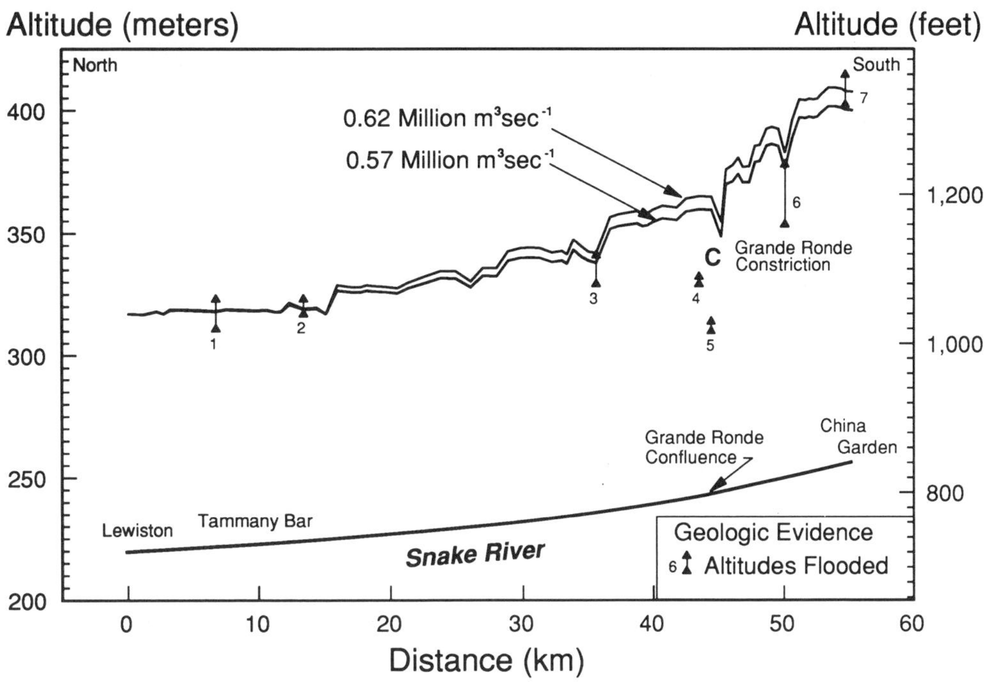

Figure 49. Modeled water-surface profiles and geologic evidence of maximum Bonneville Flood stages within the Lewiston reach.

backwater flow reconstructions is that the local energy slope can be calculated, resulting in improved estimates of local shear stress and stream power.

For evaluating erosional and depositional processes along the Bonneville Flood route, I have primarily used unit stream power as an index of geomorphic work because of its direct relationship to local rates of energy expenditure and the physical capacity of the flow to perform work. In evaluating the results, however, two caveats need to be emphasized. (1) As mentioned previously, rates of energy expenditure and shear-stress values reflect only the *potential* for geomorphic work done by the flow. For natural streams, it is very difficult to determine how efficiently the available energy is applied toward the geomorphic processes of sediment transport and erosion. A large portion of the mechanical energy of the flow is undoubtedly transformed into thermal energy associated with frictional and viscous heating. (2) The accuracy of the unit stream-power calculations depend on the quality of the flow reconstructions provided by the step-backwater modeling. In reaches of poor results, the calculated stream power values may not closely represent actual flow conditions. Even in reaches where the flow modeling results match the evidence of maximum flow stages, step-backwater calculations provide only a partial description of the actual flow conditions: flow momentum is not considered, all flow is considered to be strictly in the downstream direction, complex conditions involved with flow around bends are not adequately treated, and each cross section is considered to have a uniform energy slope for the entire width. Natural flows are highly complex and tractable treatments require multiple simplifying assumptions. Nevertheless, the step-backwater modeling provides a "first-cut" description of local flow hydraulics and their spatial variations.

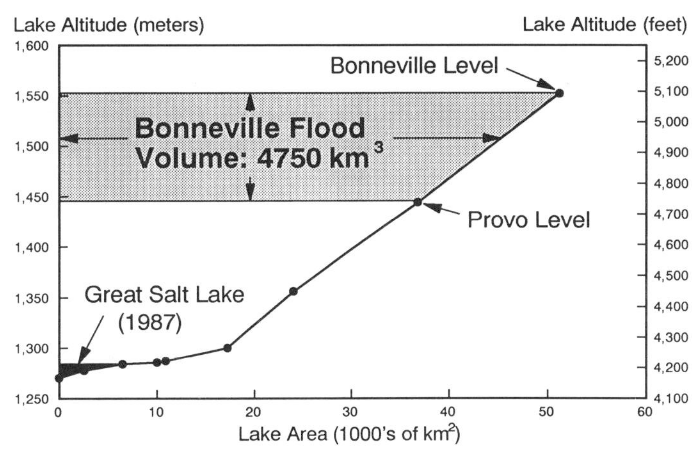

Figure 50. Lake Bonneville/Great Salt Lake hypsometry. The total flood volume was calculated with the conic equation (see Jarrett and Malde, 1987). Modified from Currey (1990).

Maximum unit stream power values within each cross section varied from less than 10 watts·m^{-2} to more than 100,000 watts·m^{-2}. The highest unit stream power values are at constric-

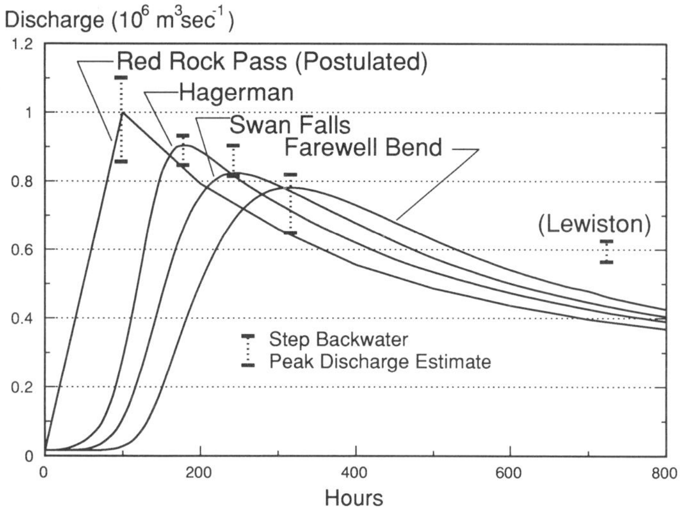

Figure 51. Hydraulically routed Bonneville Flood hydrographs for the Hagerman, Swan Falls, and Farewell Bend reaches and corresponding discharge estimates from the results of the step-backwater flow modeling. The hydrograph at Red Rock Pass was postulated and is consistent with the peak discharge at Red Rock Pass and the total flood volume; and results in downstream peak discharges consistent with the step-backwater reconstructed estimates at the downstream study reaches. Total flow duration for the postulated hydrograph at Red Rock Pass is about 7,200 hr (300 days). For comparison, also shown are the bracketing discharges determined by the step-backwater calculations for the Lewiston reach, downstream of the routed portion of the flood path.

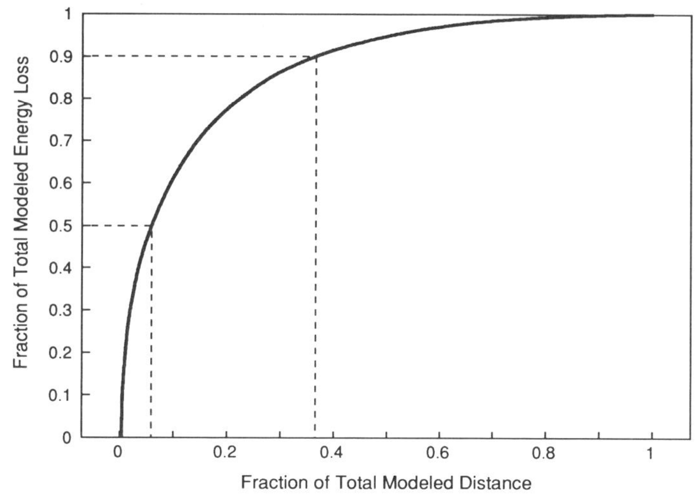

Figure 52. Cumulative distribution of energy loss with respect to flow distance for each modeled pair of cross sections within all 10 modeled reaches. Values were cumulated in order of decreasing energy slope for each cross-section pair. The dashed lines illustrate that 50% and 90% of the flow energy was expended over only about 6% and 37%, respectively, of the total modeled distance.

tions where, at several locations, maximum unit stream power values exceeded 30,000 watts·m^{-2} (Table 3). For comparison, maximum calculated unit stream-power values for historic floods in bedrock rivers range to as much as about 20,000 watts·m^{-2}. Maximum unit stream power for the largest of the Missoula Floods achieved values of 250,000 to 300,000 watts·m^{-2}. In contrast, the largest floods on the alluvial Amazon and Mississippi Rivers have produced unit stream powers of only about 12 watts·m^{-2} (Baker and Costa, 1987).

In concert with the spatially varied nature of energy loss, unit stream power magnitudes varied substantially through most reaches. For example, within the constricted portion of the Swan Falls reach, maximum cross-sectional values of unit stream power varied from more than 30,000 watts·m^{-2} to less than 10,000 watts·m^{-2} over short distances at several locations. Within the wider reaches, maximum unit stream power values were less than 1,000 watts·m^{-2} (Fig. 54). These variations apparently exerted major control over the distribution of erosional and depositional features left by the flood (Fig. 55). By mapping and measuring patterns of erosion and deposition, this relationship can be quantitatively evaluated in reaches satisfactorily modeled by the step-backwater calculations.

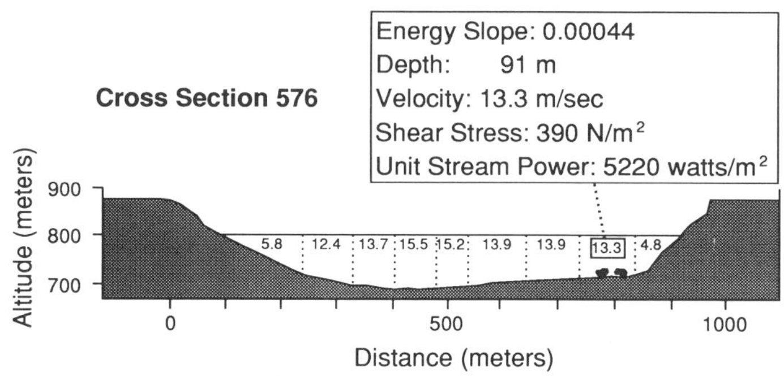

Figure 53. Calculated local hydraulic parameters from the results of the step-backwater flow reconstruction at a single cross section. This cross section is at a boulder measurement site on the Priest Ranch bar within the Swan Falls reach (Fig. 34).

DEPOSITION AND EROSION

The Bonneville Flood left a distinctive suite of erosional and depositional features (Malde, 1968, p. 13–22). Large bar complexes, fine-grained slackwater sediment, and scoured and eroded basalt can all be found at several places along the flood route. Many of the features are similar in scale and morphology to those left by the Missoula Floods in northern Idaho and eastern Washington (Fig. 1) (Bretz and others, 1956; Baker, 1973). Except for their dimensions, however, many features are also similar to those formed by floods on rivers of all scales. The point bars, pendant

TABLE 3. MAXIMUM FLOW CONDITIONS AT SOME MAJOR CONSTRICTIONS

Location	Energy Slope	Velocity ($m \cdot sec^{-1}$)	Depth (m)	Shear Stress ($N \cdot m^{-2}$)	Stream Power ($watts \cdot m^{-2}$)
Grande Ronde	0.001688	30.4	104.6	1,730	52,641
China Garden	0.001062	26.5	120.2	1,251	33,114
Pittsburg Landing	0.002235	41.2	133.8	2,931	120,863
Klopton Creek	0.000843	28.1	156.2	1,290	36,240
Farewell Bend	0.00953	23.0	105.8	987	22,728
Walters Bar	0.001621	24.0	75.4	1,198	28,710
Initial Point	0.001359	25.0	91.8	1,223	30,604
Swan Falls	0.001954	31.0	96.7	1,851	57,461
Bliss Dam	0.002397	33.3	91.9	2,159	71,825
Hagerman	0.002242	23.4	56.9	1,250	29,225
Thousand Springs	0.002120	26.6	72.0	1,494	39,699
Niagara Springs	0.002763	31.2	75.2	2,035	63,545
Rock Creek	0.002003	37.1	123.9	2,431	90,148
Hansen Bridge	0.002106	32.0	96.7	1,995	63,802
Murtaugh Reach	0.001459	30.5	120.5	1,722	52,481
Milner Dam	0.002510	18.3	39.5	972	17,800
Goose Lake	0.006350	14.4	26.2	1,700	24,414
East of Eden Potholes	0.008386	13.8	20.7	1,705	23,584
Red Hill	0.004516	15.8	49.2	2,175	34,349
Portneuf Narrows	0.001854	28.2	87.9	1,596	45,003
Marsh Creek Fan	0.002042	23.4	62.4	1,248	29,209
Red Rock Pass	0.001971	24.3	66.4	1,282	31,131

bars, and expansion bars of the Bonneville Flood are readily recognizable because of their similarity to analogous forms in modern river systems. Likewise, some of the erosional features of the Bonneville Flood may also have counterparts in modern bedrock rivers, albeit at a more reduced scale.

The erosional and depositional features of the Bonneville Flood record the physical processes driven by the local fluid dynamics. The reconstruction of flow conditions provided by the step-backwater calculations in conjunction with mapped and measured patterns of erosion and deposition can locally provide a quantitative link between hydraulic conditions at peak discharge and the resultant flood features. The general relevance of the step-backwater modeling results for understanding erosional and depositional processes is illustrated by the correlation of erosional and depositional features with the reconstructed local hydraulic conditions (Fig. 55) where it is apparent that spatial variations in flow conditions substantially influenced the distribution of flood features. In the following sections, I describe some pertinent aspects of many of the erosional and depositional flood features and show their relation to local hydraulic environments. Major emphasis is placed on the tractive deposits because of the suitability of the step-backwater modeling for quantitatively evaluating their depositional environment.

Tractive deposits and depositional processes

Most of the Bonneville Flood deposits are readily recognizable in the field by their large forms and sedimentary structures. Furthermore, they contain a much larger percentage of basalt clasts than does typical Snake River gravel, giving the flood deposits a distinctive gray appearance. Genetically, Bonneville Flood deposits are separated into two distinct classes: tractive (bedload) deposits and suspended-load deposits. The tractive deposits are composed of clasts that were transported in continual contact

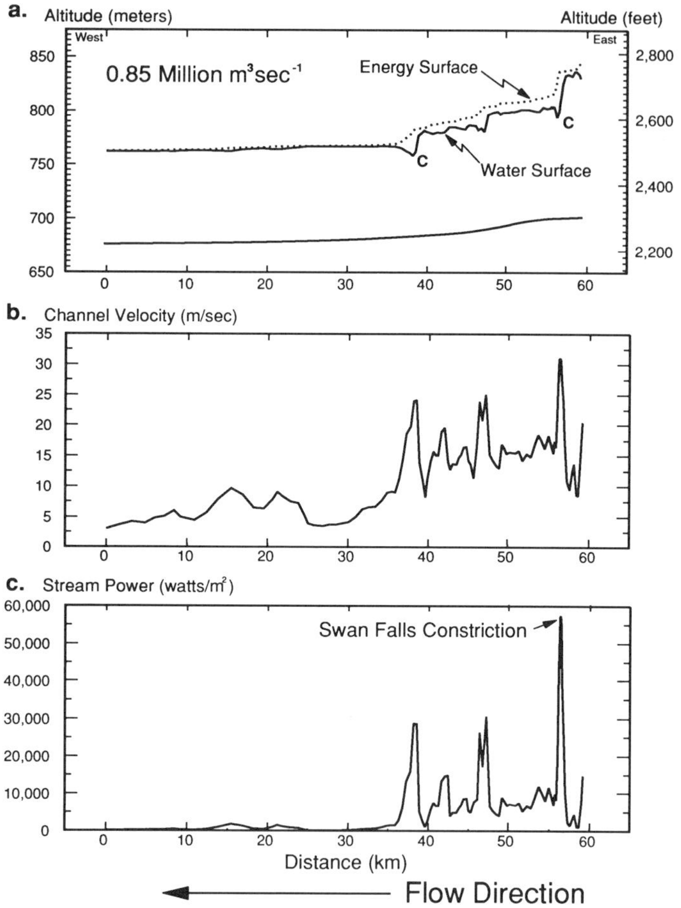

Figure 54. Hydraulic conditions along the Swan Falls reach (Fig. 34) calculated from the step-backwater flow reconstructions. Plotted values of velocity and unit stream power are maximum values within each cross section. a, Water-surface and energy-surface profiles (from Fig. 39). b, Channel velocity. c, Maximum unit stream power.

with the bed prior to deposition. Suspended-load deposits consist of clasts that were transported higher in the flow column and were supported by flow turbulence and dispersive contact with other clasts.

Expansion bars, longitudinal bars, point bars, and pendant bars are all constructional forms composed of bedload material. These features are generally near relatively steep and energetic reaches of the flood route where there was abundant material tractively transported. Tractive deposits were emplaced within the main thread of flow; consequently, the hydraulic conditions associated with deposition of tractive deposits are particularly suitable for analysis with step-backwater flow reconstructions.

Form, stratigraphy, and genesis. Expansion bars and longitudinal bars generally occupy wider portions of the flood route. The largest depositional features are the expansion bars, such as the bar complex downstream of Pocatello (Fig. 14), the bars in Hagerman Valley (Fig. 28) and in Melon Valley (Fig. 29), and Walters bar (Fig. 38). They are generally located at transitions between steep and confined segments and wide, ponded reaches. Expansion bars can be considered *boulder deltas* as defined by Elfström (1987). They are several kilometers long and contain immense volumes of material; the bar complex in Hagerman Valley consists of more than 200×10^6 m^3 of flood deposits (Malde, 1968, p. 32). The bar complex downstream of Pocatello (Fig. 14) is several times larger. Many of these bars increase in altitude in the downstream direction from an apex near present river level to several tens of meters higher at their crests (e.g., the bars at Melon Valley [Fig. 29] and Pittsburg Landing [Fig. 45]). Similar to those in the Channeled Scabland (Baker, 1973, p. 41), many of the expansion bars have marginal channels incised into their outboard margins. The largest expansion bars are extensively channeled by anastomosing channel networks that have left a surface morphology of multiple lenticular bar forms.

Longitudinal bars are narrow, streamlined sediment accumulations that flank and parallel the main channel. They were generally deposited downstream of constrictions in locally wider canyon segments (e.g., the bar near Arimo [Fig. 16] and the Initial Point bar [Figs. 36, 37]). They are smaller than expansion bars and have axial lengths as long as 3 km. There is commonly a marginal channel between the bar crest and the flow margin.

The point bars left by the Bonneville Flood have similar morphologies to those in modern fluvial systems. They are deposited along the insides of channel bends and commonly have a surface that descends toward the main channel. Some of them apparently resulted from decreased flow velocity and deposition associated with flow along the inside of a channel bend (e.g., the point bars near Lewiston [Fig. 48]). Many of them, however, may also be a consequence of coeval channel migration due to erosion along the outside bend. The point bars at Swiss Valley (Fig. 33), upstream of the Hagerman bar, and upstream of Walters bar are probably the result of such a process. In these locations, flow was forced around abrupt bends where the outside valley wall consisted of soft alluvial deposits. Furthermore, as discussed in a previous section, results from the flow reconstructions are consistent with channel enlargement at some of these locations during passage of the flood.

Pendant bars (Malde, 1968, p. 44; Baker, 1973, p. 35–36) are sediment accumulations in the lee of resistant protrusions into the flow. They are generally narrow and sharp crested, with maximum lengths of 1 to 2 km (e.g., the pendant bar downstream of White Butte on Walters bar [Fig. 38]). Commonly, there is a depression surrounding the flow protrusion that separates it by a short distance from the pendant bar. This erosional moat is generally deeper in front of the flow protrusion and is probably the result of local scour associated with the intensified turbulence set up by the flow perturbation.

Where exposed, the internal stratigraphy of the tractive deposits is remarkably similar. Downstream-directed foresets are capped by 1 to 3 m of horizontally bedded and distinctly coarser clasts (Fig. 56). Measured foreset inclinations range between 15° and 34° and are generally oriented within 30° of the inferred

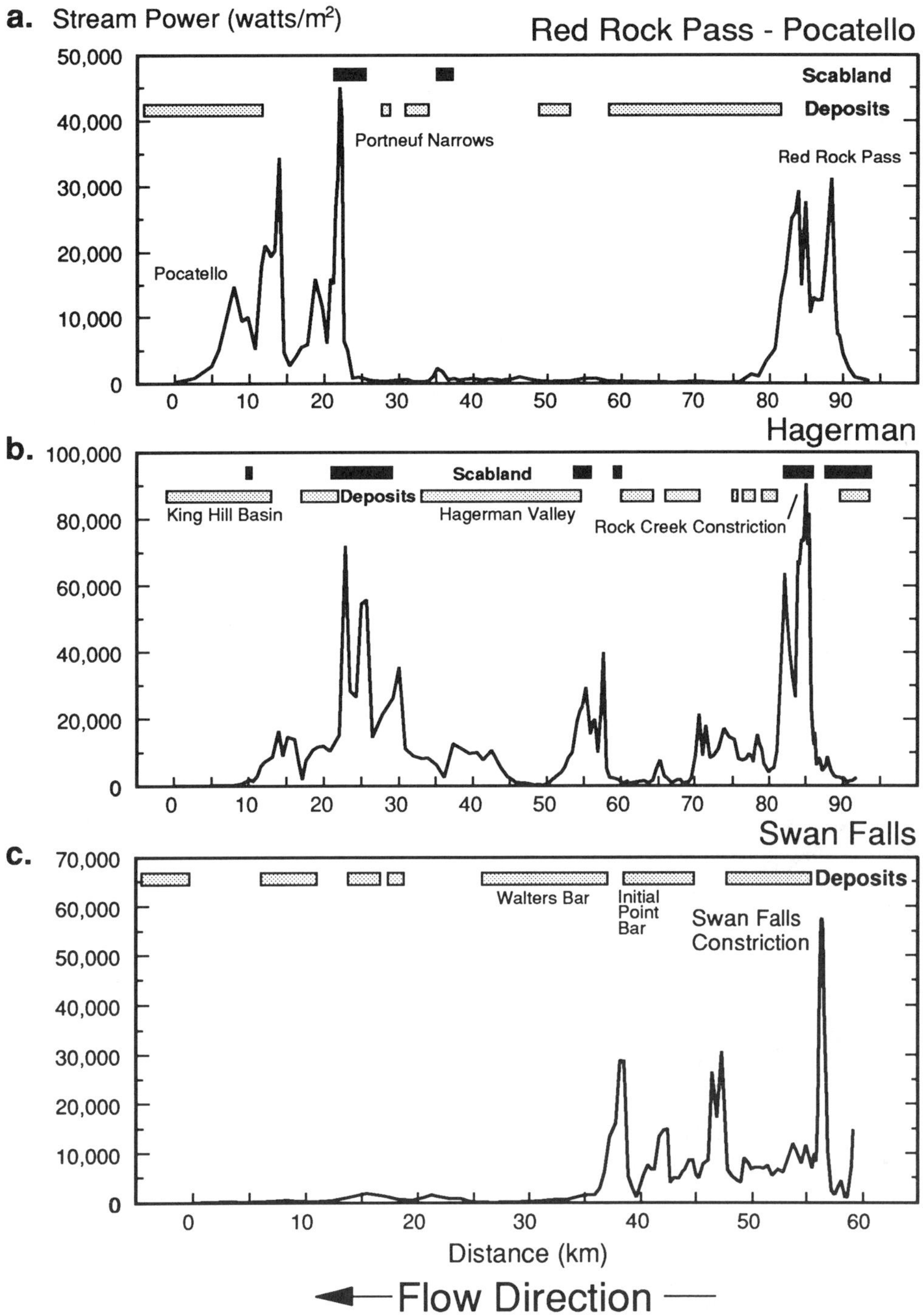

Figure 55. Maximum unit stream power values for three of the study reaches. The stream power magnitudes and their spatial variation are highly correlated to the distribution of erosional and depositional features. Indicated areas of erosion are restricted to butte-and-basin scabland topography and areas where portions of basalt flows have been completely removed. Shown areas of deposition include only tractive deposits (expansion bars and longitudinal bars). Erosion of scabland topography was generally restricted to short reaches that had unit stream power values exceeding 20,000 watts·m^{-2}. a, Red Rock Pass to Pocatello. b, Hagerman reach. c, Swan Falls reach.

local flow direction (as estimated from the orientation of the channel and local flood features). The foresets consist of 0.25- to 1.5 m-thick beds of coarser material alternating with finer beds. Sorting within the foresets is generally poor, and all transported grain sizes, from silt to boulders, are present. At wider bars, where there are exposures perpendicular to the flow direction, there are trough-and-fill structures with dimensions of tens of meters. The upper 1 to 3 m of the tractive bars generally consist of better sorted and coarser material deposited in horizontal beds that disconformably overlie the foresets (Fig. 56). The clasts within this horizontally bedded unit are of sizes similar to the maximum-sized clasts deposited within the foresets. Maximum clast size in the upper layer and within the foresets invariably decreases downstream within individual bars.

Figure 56. Exposure of tractive bar stratigraphy in a gravel pit at the downstream end of the expansion bar in Melon Valley (Fig. 29). Flow direction was left to right. Crudely bedded downstream dipping foresets are overlain by a 2-m-thick subhorizontal layer composed primarily of clasts from the coarsest size fraction in the deposit. At this site, the clast sizes in the subhorizontal layer appear to decrease upward. The Bonneville Flood deposits are capped by about 1 m of postflood loess.

The stratigraphy indicates that the bars composed of tractive deposits form by downstream growth rather than vertical accretion. The model for their formation proposed in Figure 57 is similar to that presented by Baker (1973, p. 36). These bars grow by accreting foresets on their downstream termini from bedload being transported across the bar surfaces. The trough-and-fill structures exposed in some of the wider bars indicate that they may begin to grow downstream as a series of fingers with the interfluves being filled later in the flow; the two series of bars at the downstream end of the Eden channel (Fig. 21) may represent this style of bar growth arrested early in the process.

The poor bedding within the foresets may reflect the abrupt change in transport conditions at the downstream end of the bar. As bedload clasts moving along the top of the bar reach the bar terminus, they enter a zone of flow separation (Fig. 57), and many of the clasts that are trapped in the reverse current in the lee of the bar terminus drop out of the flow because of the substantially reduced local flow velocity. Continued rapid deposition traps these clasts in unorganized depositional positions. Additionally, because of the extreme turbulence in this flow separation zone, the bar termini are probably major launching zones of clasts that escape the reverse circulation at the bottom of the flow and become part of the suspended load. The crude bedding defining the foresets could reflect surges in the flow, or the passage of bedforms or bedload sheets across the bar surface.

The upper 1 to 3 m of horizontally bedded coarser deposits are probably bedload clasts that were traveling along the top of the bars. Because of the simpler hydraulic conditions and the surface stability, the bar surfaces have probably been winnowed of most of the finer component of the clasts, leaving an armored surface of oriented cobbles and boulders. Armoring would have occurred continuously at a particular location from the time of bar growth past that position until the flood stage receded below the altitude required to maintain downstream-directed flow over the bar. As is discussed in detail later, the clasts that were deposited in this armored layer probably rolled to their present position, and their dimensions closely relate to local flow competence during the time of their deposition.

Boulder shape and fabric. A striking aspect of the tractive deposits is the strong fabric shown by the boulders mantling the bar surfaces (Malde, 1968, p. 32). Bonneville Flood boulders of all sizes are preferentially oriented and are well rounded. There have been extensive field and experimental studies of particle form, roundness, and depositional orientation with regard to sand and pebble deposits (see Johansson [1976] for a review). There are few data, however, for documenting fabric development in large, boulder-sized bedload clasts with the exception of the studies by Elfström (1983, 1987) of Pleistocene flood deposits in northern Sweden. The Bonneville Flood boulders have form, fabric, and roundness characteristics very similar to those developed in tractively transported pebbles and cobbles deposited by smaller fluvial systems and are suggestive of similar transport and depositional mechanisms.

Boulder measurements at sites along the flood route (Fig.

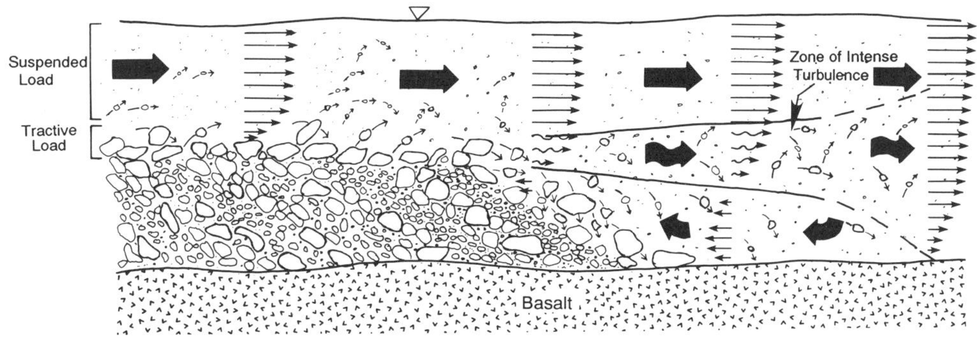

Figure 57. Schematic representation of the flow and sediment transport conditions associated with growth of tractive bars during the Bonneville Flood. Wide arrows represent the general flow direction. The thin arrows schematically represent relative velocity magnitudes and the degree of turbulence. Some aspects of the postulated flow conditions are taken from Jopling (1963).

58) allow for description of some aspects of the depositional fabric. Clast orientation was measured for the boulders that were apparently in original depositional positions; roundness was measured for almost all of the measured clasts with the visual guide of Powers (1953) (Fig. 59). Measurements from deposits on the Snake River Plain are most appropriate for analysis because almost all of the boulders are basalt (providing a fairly homogenous sample), and most of the clasts are derived from nonfluvial sources.[5]

[5]Large basalt boulders are a minor component of Holocene and Pleistocene Snake River deposits, and the basalt clasts in the Bonneville Flood deposits were most likely derived from talus flanking the canyon walls or from erosion of basalt flows during the flood.

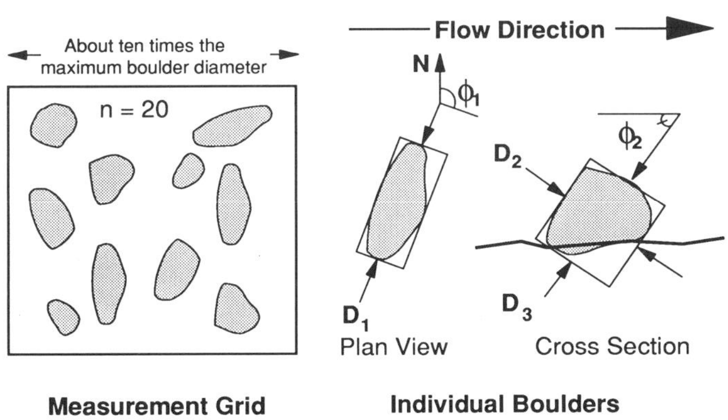

Figure 58. Methods adopted for measuring boulders that are in original depositional positions. At each measurement site, 20 boulders were measured within an area scaled to boulder size. For the individual boulders, I measured the lengths of the three mutually perpendicular axes (D_1, D_2, D_3) that defined the volumetrically smallest rectangular solid that was estimated to completely enclose the boulder. Orientation measurements [compass direction (ϕ_1) and inclination (ϕ_2)] were made for the most upstream-dipping face of the visualized solid. For boulders that were partially buried, minimum measurements were recorded. The boulder measurements are summarized in Appendix 3. Complete measurement data are available from the author.

Sneed and Folk (1958) proposed guidelines for describing particle form that have been adopted for classifying the Bonneville Flood boulders (Fig. 60). The measured boulders are fairly evenly distributed among the classes in the upper part of the form triangle, with the compact-bladed and bladed classes containing the largest percentages (Fig. 60). Some of the shapes are obviously inherited from irregularities in the source basalt flows. Many of the large rod-shaped clasts are transported basalt cooling columns that have retained their polygonal cross sections, and many of the large plate-shaped boulders are bounded by surfaces that were contacts or irregularities within the basalt flow stratigraphy.

Boulder orientation. Commonly evaluated orientation features are clast inclination (the upstream dip or imbrication angle) and long-axis orientation with respect to flow direction. Bonneville Flood boulder imbrication is most generally of the *isolate* (Johansson, 1976) or *scattered* (Leopold and others, 1964, p. 210) types where the largest clasts have a preferred upstream inclination but are not in direct contact with each other (they are not shingled but are oriented within a matrix of smaller clasts). There are also several examples of "lodgement orientation—the approximately coplanar sequence or chains of three to five particles long" (Sedimentary Petrology Seminar, 1965) (Fig. 61). Figure 62a portrays the inclination angles for all of the Snake River Plain boulders for which there are orientation data, revealing an overall concentration of inclination angles between 10° and 40° in a manner similar to most pebble deposits (Blatt and others, 1980, p. 123). Separating the boulder measurements on the basis of boulder shape, it is apparent that the platy boulders are more strongly inclined at lower inclination angles (Fig. 62b). The modal inclination angle class of the platy boulders is 15° to 20° compared to 35° to 40° for the compact boulders. This is expected hydrodynamically for bedload particles tractively transported along the bed and has been documented in flume studies with pebbles (Johansson, 1976).

The relation between the measured orientation of the long axis of each boulder to flow direction was determined by independently estimating local flow direction from the local

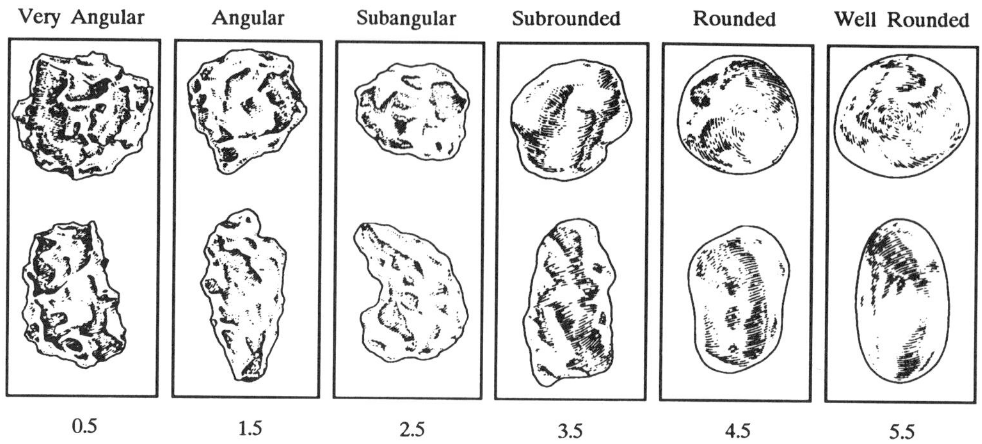

Figure 59. Roundness measurement guide of Powers (1953 modified from Pettijohn and others, 1972, p. 586). Roundness values were estimated in the field to a precision of about ±0.1. For the 1998 boulders and cobbles measured along the flood route, mean clast roundness was 3.8 (standard deviation = 0.6). For 1,213 boulders within the Swan Falls and Hagerman reaches, mean roundness was 4.05 (standard deviation = 0.55).

channel geometry and the orientation of bar forms. The dominant orientation of the long axes is perpendicular to the local flow direction, with a weak secondary preferred orientation parallel to flow (Fig. 63a). The elongate clasts are more strongly oriented (Fig. 63b).

Clast orientation has strong influence on the clast's susceptibility for entrainment. Elongate particles transverse to the flow direction are the most susceptible to entrainment (Johansson, 1976, p. 208), and their presence implies that local hydraulic conditions were too weak to sustain continued transport. In contrast, elongate clasts deposited in highly energetic environments—that is, flow normally sufficient to maintain transport—are only stable if emplaced in a position parallel to flow because of the smaller area projecting into the flow. From flume studies with pebbles, Johansson (1976, p. 229–249) concluded that for a given pebble-size distribution, depositional clast positions with long axes orineted perpendicular to flow direction are preferred with minimum transport rates; with more vigorous flow conditions, a longitudinal orientation is preferred. The perpendicular orientation of the long axes of the Bonneville Flood boulders indicates that they rolled to their present positions, probably within a flow environment that approached minimum sustaining transport conditions.

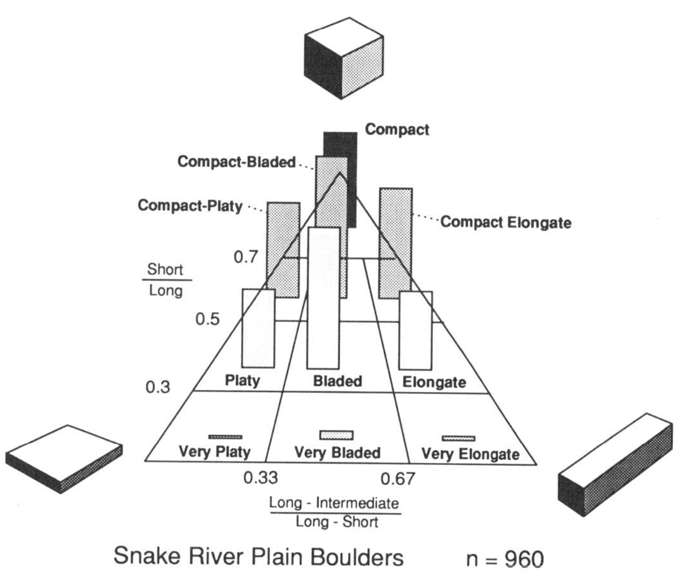

Figure 60. Clast form classification scheme of Sneed and Folk (1958) with distribution of tractively deposited Bonneville boulders measured on the Snake River Plain.

Clast rounding. Russell (1902, plate XXV-b) noted the "water-worn" nature of the boulder deposits near Hagerman despite not describing their origin. The well-rounded Bonneville boulders are distinctive along the entire flood route, and for the boulders on the Snake River Plain, the rounding is noteworthy because of their large sizes and short transport distances. Average particle roundness was 4.0 (see Fig. 59 for the roundness measurement index) for the 1,493 boulders measured on the Snake River Plain. Directly determining transport distances for individual clasts is difficult on the Snake River Plain because of the variety of basalt flows that contributed clasts to the deposits. However, the Swan Falls study reach provides an opportunity to estimate transport distance for clasts deposited downstream of the Swan Falls constriction. Flow upstream of the constriction was hydraulically ponded, and there is no evidence of transport of clasts larger than cobbles from sources upstream of the constriction to downstream locations. Therefore, assuming that all of the boulders deposited downstream of the constriction were derived from the constriction or the intervening canyon, their distance below the constriction is the maximum distance the boulders could have

Figure 61. Photograph of an imbricated sequence of boulders deposited on the Walters bar within the Swan Falls reach (Figs. 34, 38). Flow was from right to left.

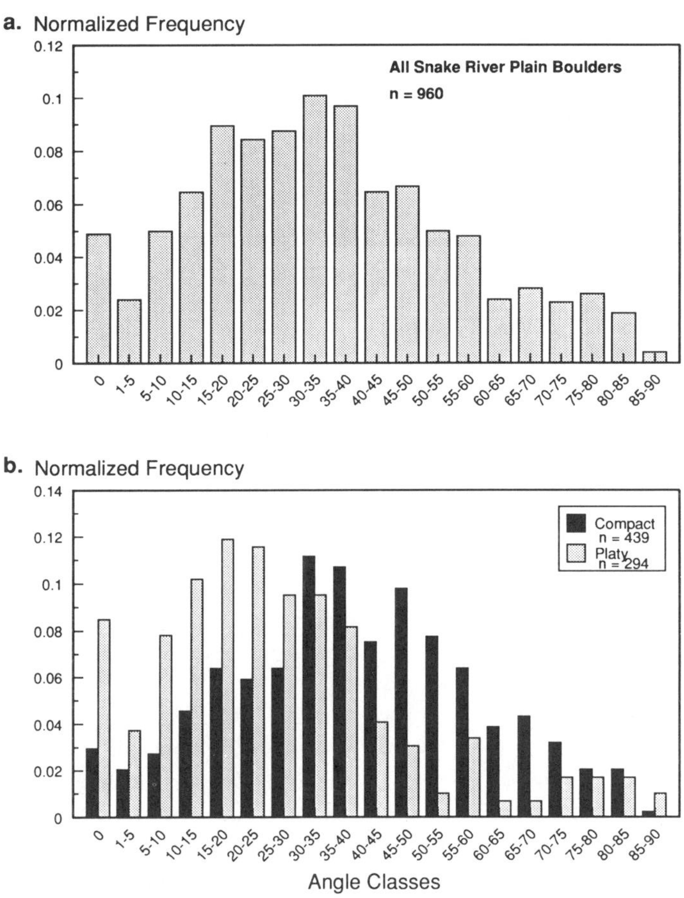

Figure 62. Boulder inclination measurements for Snake River Plain boulders in original depositional positions. a, All Snake River Plain boulders. b, Measurements discriminated on the basis of boulder shape. The platy boulders include boulders from the platy, bladed, very platy, and very bladed classes of Figure 60. The compact boulders are those in the compact, compact-bladed, and compact-elongate classes. The platy boulders were deposited with shallower inclination angles than boulders of more compact forms.

traveled. Average boulder roundness plotted against this distance constrains the degree of rounding with distance (Fig. 64). Ultimate roundness values for the basalt clasts were achieved within maximum transport distances of 10 to 25 km. Trends within individual bar complexes are suggestive of even greater rounding rates. Although this analysis is not precise, the results are in agreement with Malde's (1968, p. 14) qualitative assessment based on his recognition of source areas. Furthermore, these results are consistent with numerous studies evaluating pebble rounding (see Mills [1979] and Sneed and Folk [1958, p. 126–129]).

The primary conclusion from the shape and fabric data is that the Bonneville Flood tractive deposits are remarkably similar to those measured for pebbles and cobbles in flumes and small streams. This implies, despite being difficult to comprehend, that the 0.5- to 11-m Bonneville Flood boulders were transported and deposited by processes analogous to those observed with trac-

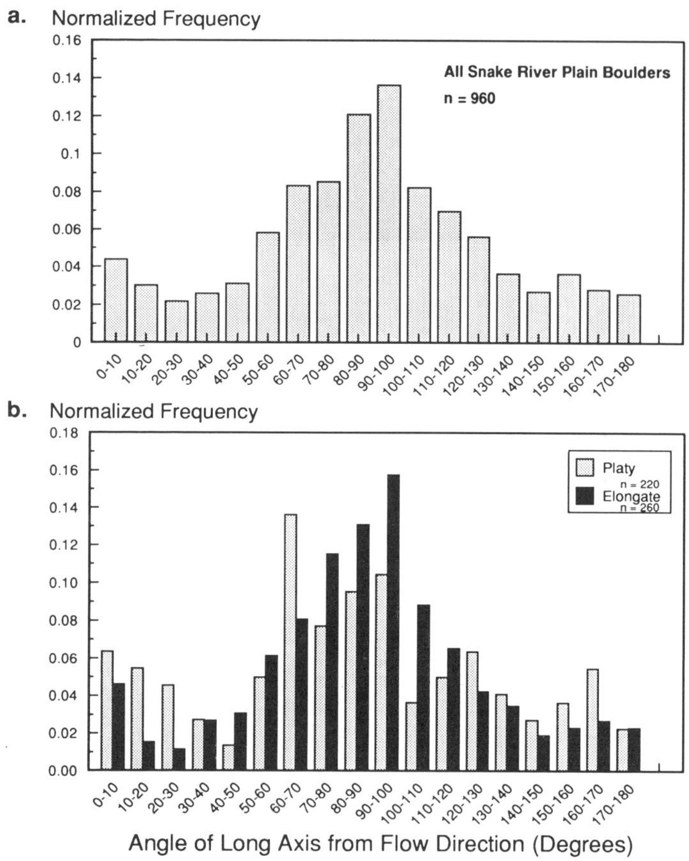

Figure 63. Long axis orientation with respect to flow direction. a, All measured in-place Snake River Plain boulders. b, Measurements discriminated on the basis of boulder shape. The platy boulders include compact platy, platy, and very platy classes of Figure 60. The elongate boulders include the compact-elongate, elongate, and very elongate shape classes. More than 50% of the long axes of elongate boulders are oriented within 20% perpendicular to the local flow direction.

tively transported pebbles and cobbles in more typical-sized fluvial systems. Moreover, the bar forms and the fabric data indicate that the clasts mantling the surfaces of the tractive deposits were probably emplaced under fairly stable flow conditions at or near the limit of sustained transport.

Analysis of boulder transport and deposition. A major emphasis of this study was to use the results of the step-backwater calculations to evaluate boulder transport and deposition. Coarse particle transport and flow competence have been studied by geologists and engineers since early in this century (e.g., the analyses by Gilbert [1914] and Rubey [1938]) to understand the ability of fluid flow to move bedload. The goal of many of these studies, both theoretical and empirical, has been to develop a relationship between some index of flow strength (such as velocity, shear stress, or stream power) and the largest clasts transported by the flow. From the geologist's perspective, such a relationship can aid paleoenvironmental and paleohydraulic analyses by providing quantitative information regarding past flow conditions. Relationships developed by Baker and Ritter (1975), Costa (1983), and Williams (1983) have been applied for these types of paleohydrologic analyses.

Existing relationships and analyses are imprecise at best, especially for particle diameters exceeding a few centimeters. Several factors contribute to this. A major obstacle in achieving a theoretical understanding of coarse particle transport is the complex and turbulent flow structure at the fluid-bed interface. The interaction between a particle and the flow depends on many aspects of the channel bottom and the nature of the sediment load. Factors such as particle-size distribution, particle shape, and grain packing all affect the potential for transport, as well as the local flow structure (Baker and Ritter, 1975; Church, 1978; Costa, 1983; Williams, 1983; Komar, 1987; Komar and Li, 1988; Komar and Carling, 1991).

Additionally, the paucity of consistently obtained data hinders both empirical and theoretical analyses. The compilation of data by Komar (1987) (Fig. 65) exemplifies some of these limitations: the data are from many different sources, representing different scales of flows and particle-size distributions. Furthermore, different properties of the studied systems were measured; some of the studies reported the single largest clast, or the 5 or 10 largest clasts, that were moved. Some reported a percentile of the grain-size population, such as the largest 10%. In some cases particle mass was reported rather than particle diameter. A critical problem for the analysis of large and geomorphically effective flows is the sparse data for clasts larger than 100 cm. Also, workers have measured different aspects of the transporting flows with a variety of measurement techniques. Most studies report direct measurements of velocity, but many were made at different levels in the flow.

Some of the largest clasts in the compilation of Figure 65 are those measured by Scott and Gravlee (1968) associated with a dam failure on the Rubicon River, California. Here, the flow measurements were made indirectly and the tractive forces were calculated using the channel slope as a proxy for energy slope. As stated earlier, this substitution is not always appropriate (O'Brien and Rindlaub, 1934). In fact, almost all of the unit stream power and shear-stress values used in these types of compilations are derived with values of channel or water-surface slopes instead of energy slope. Clearly, as many of these workers fully appreciated, this empirical approach has suffered from a "few apple crates and a few orange trees" (Mackin, 1963, p. 141) getting onto the graph. Williams (1983), in a similar type of compilation using much of the same data, noted a substantial overlap in velocity, shear stress, and stream power values associated with transport and nontransport for a given size particle, and stated (p. 228) that "there is no general and unique threshold flow for a given [transported] grain size."

The goal of most of these competence analyses was to define the *critical tractive force* or threshold conditions for *initiating* particle motion. Nevertheless, some of the data defining these relationships were associated with observed or measured conditions of particles *in motion,* and hydraulic conditions indirectly calculated in association with *depositional* positions of clasts. Be-

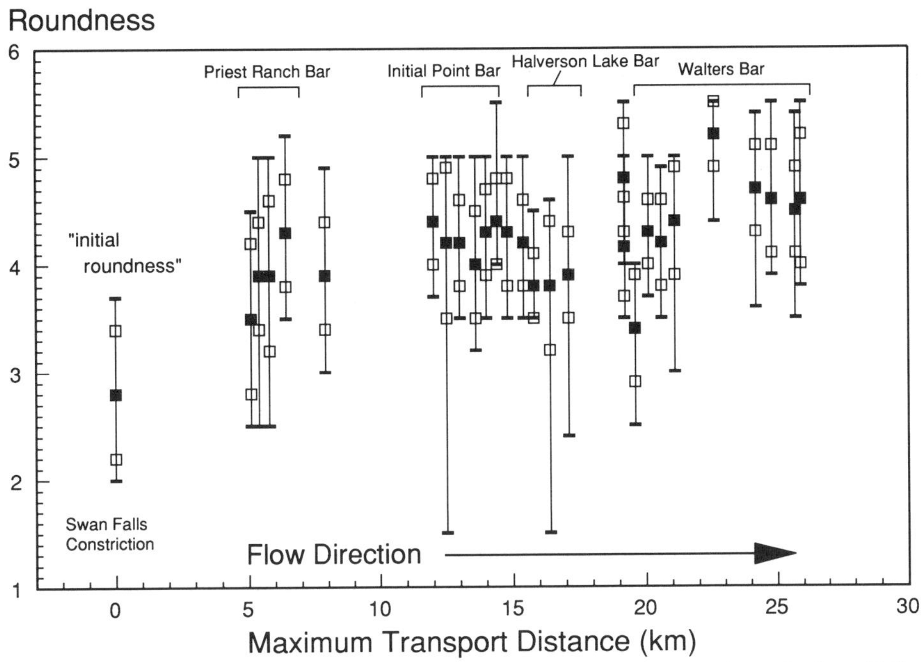

Figure 64. Boulder roundness (Fig. 59) with respect to distance downstream from the Swan Falls constriction. Mean roundness values, standard deviations, and measurement ranges are plotted for the 20 largest boulders at each measurement site. Roundness values generally increase within the Swan Falls reach (Fig. 34) and within the individual bar complexes. A new source of bedload clasts from talus accumulated along the canyon wall may have contributed to the small downward step in roundness values at the upstream end of the Halverson Lake bar. An estimate for "initial roundness" at the Swan Falls constriction was obtained from the site of lowest average roundness value measured for Bonneville Flood boulders on the Snake River Plain. This site was in the Hagerman reach and consists of boulders that were probably transported only a few meters from a nearby basalt ledge. For comparison, almost all talus blocks that flank the canyon walls have roundness values less than 3.0.

sides resulting in different types of data being used to establish transport relations, an important issue is raised; the processes and conditions associated with initial particle motion and particle deposition are fundamentally different and need to be more fully discriminated in discussions of flow competence. There have been classic theoretical and empirical treatments of initial particle motion (Gilbert, 1914; Rubey, 1938; Helley, 1969). However, the findings from these studies only partly relate to understanding processes and conditions for particle deposition. Moreover, a flood deposit is the physical record of deposition, not of initial motion (Hjulström, 1935; Church, 1978). There are many mechanisms that can introduce bedload clasts into a flow and clasts can travel long distances before being deposited. These factors limit the utility of initial motion studies for paleohydrologic reconstructions based on measurement of clasts in deposits.

Conceptual framework for the analysis of the Bonneville Flood boulders. Boulder deposits left by the Bonneville Flood provided an opportunity to evaluate boulder deposition with a large and consistently derived data set. There were abundant coarse clasts deposited by the flood with measured diameters ranging from less than 10 cm to more than 10 m. Moreover, the Bonneville boulders can be appropriately evaluated within the conceptual framework that recognizes that they represent depositional conditions; that is, the hydraulic conditions at depositional sites are postulated to be the limiting conditions for sustained clast transport rather than a critical entrainment threshold for the maximum clast sizes present. In other words, a bedload clast in transport (it does not matter how or from where that particle initially moved), will continue to move until the transporting ability (the local velocity, shear stress, or stream power) of the flow diminishes (temporally or spatially) below that necessary to maintain clast movement. The systematic downstream decrease in maximum boulder size on tractive bar surfaces indicates that this limit of sustained clast transport is at least partly dependent on clast size.

Several assumptions are required to relate the results of the step-backwater modeling to boulder size in order to define this relationship. Foremost, the results depend on the quality of the modeling. For this reason, the primary emphasis is on the Swan Falls and Hagerman reaches where there are abundant boulders and where the flow reconstructions are thought to accurately portray peak flow conditions. Another required assumption is that all the measured boulders were indeed transported. Almost every measurement location was on the surface of a construc-

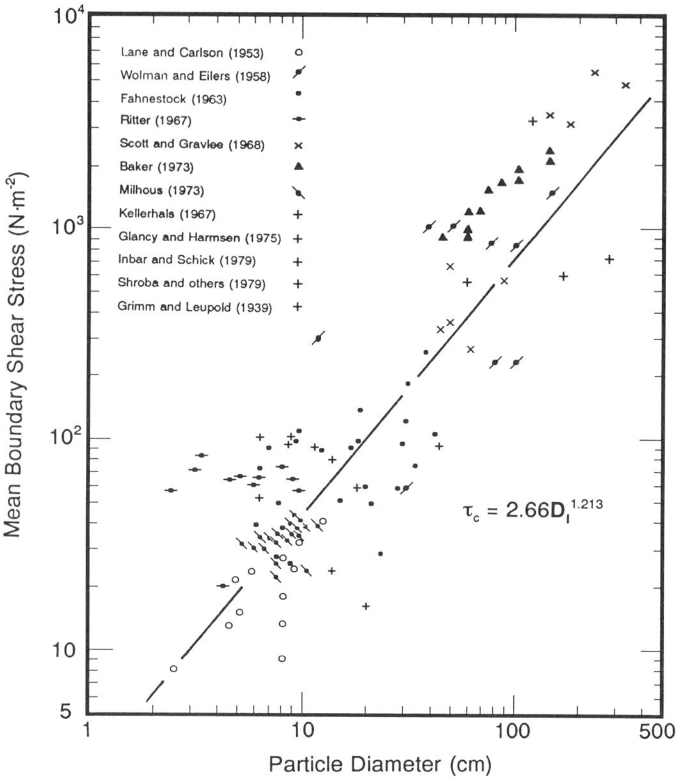

Figure 65. Compilation of coarse-particle flow competence data for empirically determining the critical boundary shear stress for clast entrainment. The data were compiled by Komar (1987) and the line was determined by Costa (1983) from a slightly different set of data (modified from Komar, 1987).

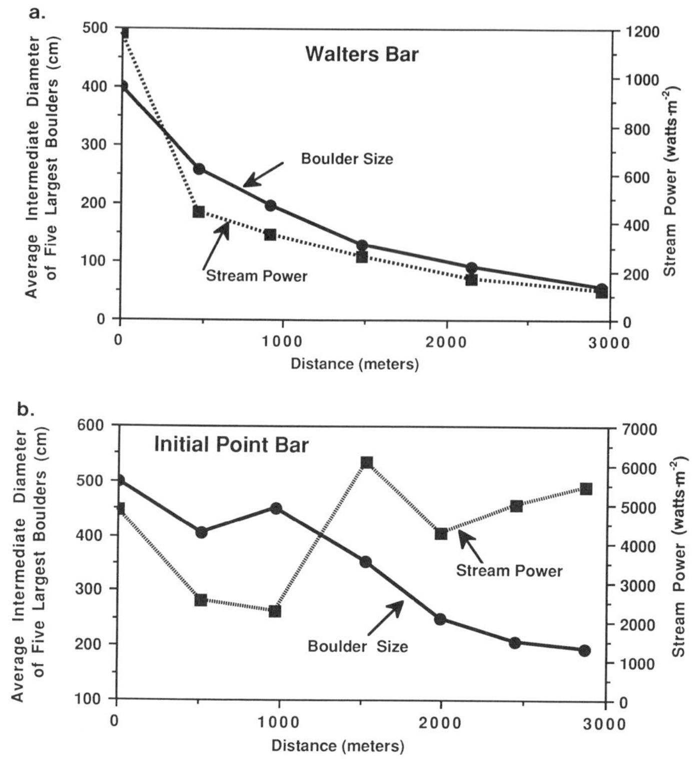

Figure 66. Local calculated unit stream power values and boulder sizes for a series of measurements on two bars within the Swan Falls reach. a, For Walters bar (Fig. 38), local stream power and boulder size decline systematically in a manner that suggests selective deposition associated with hydraulic conditions of minimum sustained transport. b, For the Initial Point bar (Fig. 37), boulder size also decreases in the downstream direction; however, the local calculated stream power does not monotonically decrease in concert with clast size. Maximum boulder size will not increase in reaches of increasing calculated stream power at locations where there are no intervening sources providing large enough clasts to define limiting transport conditions. Consequently, for boulder measurement sites at the downstream end of Initial Point bar, there are larger calculated stream-power values than those that reflect limiting transport conditions for the measured boulders. Furthermore, in comparison with measurements shown in a, stream power values associated with deposition of the largest boulders on the Initial Point bar are substantially higher than those at Walters bar. It is probable that the entire Initial Point bar, or at least the present surface, was deposited later in the flow, after peak discharge, when stream power values were lower. In this case, the calculated stream power values for the measurement sites on the Initial Point bar would be larger than the actual limiting stream power conditions for the measured boulders. Importantly, for this analysis, both types of errors lead to overestimation of local stream power values associated with limiting transport conditions.

tional bar form, positions requiring that the boulders were transported some distance. At the few sites of boulder accumulations that were not on bar surfaces, boulder measurements were not included in this part of the study unless there was convincing evidence of transport, such as rounding, imbrication, and some distance between the boulder locations and potential sources.

Also implicit in relating the hydraulic conditions of the peak discharge to boulder sizes is that all the boulders were deposited at or near the maximum discharge. This is undoubtedly incorrect; many of the measured boulders on bar surfaces were probably deposited during lower discharges after peak discharge. Consequently, some boulders were probably deposited at stream power magnitudes lower than those calculated in this analysis (unit stream power values at a site almost invariably increase with discharge). Another important assumption is that there were sufficient boulders of the appropriate sizes entrained in the flow such that the maximum possible boulder size, representing limiting transport conditions, was deposited at each measurement site. However, at measurement sites far removed from areas of entrainment, or at sites downstream of reaches with low transport capability, larger clasts may have been absent. This would also result in smaller-than-expected clast sizes associated with the hydraulic conditions of peak discharge at some sites. Figure 66 illustrates a situation where both of these cases for potential errors in determining the flow strength necessary for sustained motion may have occurred. Importantly, though, in both types of situations, these two potential errors result in *greater* estimated stream powers associated with maximum particle size at a particular site.

Because of this potential for local overestimation of the hydraulic parameters associated with deposition, the primary goal was to define the minimum stream power value associated with the deposition of a particular size clast. This is appropriate if the modeling results are accurate and all the measured clasts have indeed been transported, because, within the conceptual framework outlined above, it is impossible for clasts larger than those that closely represent the ability of the flow to sustain clast transport, at peak discharge, to otherwise get to a site. Hence, the conditions associated with the limit of sustained transport, or selective deposition, should be defined by the lower limit of the measurements relating particle size to local hydraulic conditions. Additionally, the modeling strategy of determining the *minimum* peak discharge for each reach further ensures that the resulting relationship defines the *minimum* possible hydraulic conditions for sustained transport at depositional sites for the largest clasts at each site.

Boulder measurements. Boulders were measured at 127 sites within 8 of the 10 modeled reaches. Of these sites, 75 were in the modeled portions of the Hagerman and Swan Falls reaches. Measurement locations were restricted to tractive bar deposits, primarily expansion and longitudinal bars, because of the suitability of the step-backwater modeling for reconstructing flow conditions in these environments. Measurement locations were selected partly on the basis of exposure; however, in order to define most narrowly the minimum sustaining transport conditions, it was also the intent to include the largest boulders present in the analysis. For large bars, I made measurements at several locations in order to define downstream relationships (e.g., Fig. 38).

Two types of measuring methods were adopted, depending on clast exposure. At locations where boulders were in their original depositional position and well exposed, measurement areas were laid out that approximately encompassed the area of a square with sides 10 times the long diameter of the largest boulders (Fig. 58). This simply served as a guideline for determining a reasonable, but scaled, area to confine the search for the largest boulders. Within each measurement area, the 20 largest appearing boulders (in terms of mass) were measured. Ideally, a complete sample of the pavement layer would have been measured, but this was impractical due to the size of the boulders and the 1-m blanket of loess that has accumulated on most of the bar surfaces. For each boulder, three axes were measured, as well as the orientation and inclination of the axis that was most parallel to the presumed flow direction and was inclined in the upstream direction (Fig. 58). Where axes were not completely exposed, generally because of postflood loess accumulation, minimum axis measurements were recorded.

Clast orientations were not measured at sites where the boulders were not in original depositional positions; only axis lengths and boulder roundness were measured for the 20 largest. Almost all clasts with diameters less than 100 cm were measured in this manner from spoil piles at gravel pits or from other disturbed areas.

Boulder size and local unit stream power values. The 20 largest appearing boulders were measured at each site to ensure that the 5 largest had been included in the measurement set (measuring the 10 apparent largest boulders would probably have sufficed). The five largest at each site were determined on the basis of estimates of boulder mass (for an ellipsoid of equivalent axes) and then the average intermediate diameter for these five boulders was compared with the local hydraulic variables determined from the step-backwater flow reconstructions (as in Fig. 53). These data include sites of boulder accumulations where average intermediate diameters range from 10 to 624 cm. The largest measured long diameter of a transported boulder was 1,120 cm.

When plotted on log-log paper, the data from the Hagerman and Swan Falls reaches define a lower envelope curve, drawn by eye, that represents minimum unit stream power values for transport of the Bonneville Flood boulders (Fig. 67a). The same line adequately bounds almost *all* of the Bonneville Flood boulder data except for a group of measurements at Pittsburg Landing in Hells Canyon (Fig. 67b). At Pittsburg Landing, as elsewhere in Hells Canyon, the modeled hydraulic conditions apparently have little relation to actual flow conditions during deposition of the large boulders on this bar. As described above, points plotting above the line may represent sites where measured clasts were deposited at discharges less than peak discharge, or sites where boulders of sufficient size to represent flow competence at peak discharge were not entrained in the flow.

An unexpected result of this analysis is the sharply defined *upper* limit to the unit stream power magnitudes at sites of deposition (Fig. 67). Deposition was apparently restricted to locations where, at peak discharge, unit stream powers were less than 6,000 watts·m^{-2}. Because much greater unit stream power values were achieved in the flood (as much as 100,000 watts·m^{-2}), this limit, at least with respect the largest particles that were deposited, may reflect an upper limit of the clast sizes that were entrained in the flow. That is, the maximum deposited particle sizes may reflect the critical conditions upstream for clast *entrainment.* For example, in the Swan Falls reach, the maximum average intermediate diameter for a measurement location is 624 cm. The maximum unit stream power value upstream of the measurement location was at the Swan Falls constriction where there was a maximum stream power magnitude of 57,500 watts·m^{-2}. Applying Costa's (1983) relationships between particle diameter and entrainment stream power (and extrapolating somewhat beyond its data base), the minimum required stream power to entrain a 6.24-m boulder is between 22,000 and 75,000 watts·m^{-2}, a range encompassing the maximum stream power in the constriction. Of course, the maximum particle size deposited may also reflect structural characteristics of the source rocks such as jointing in the basalt flows or the maximum size of talus blocks flanking the canyon walls. Nevertheless, this upper limit is intriguing and may indicate that critical flow conditions for large clast entrainment are several times greater than the conditions at sites of deposition.

Comparison with other coarse particle transport relationships. Empirical sediment transport relationships are commonly based on power law regressions (e.g., Baker and Ritter, 1975;

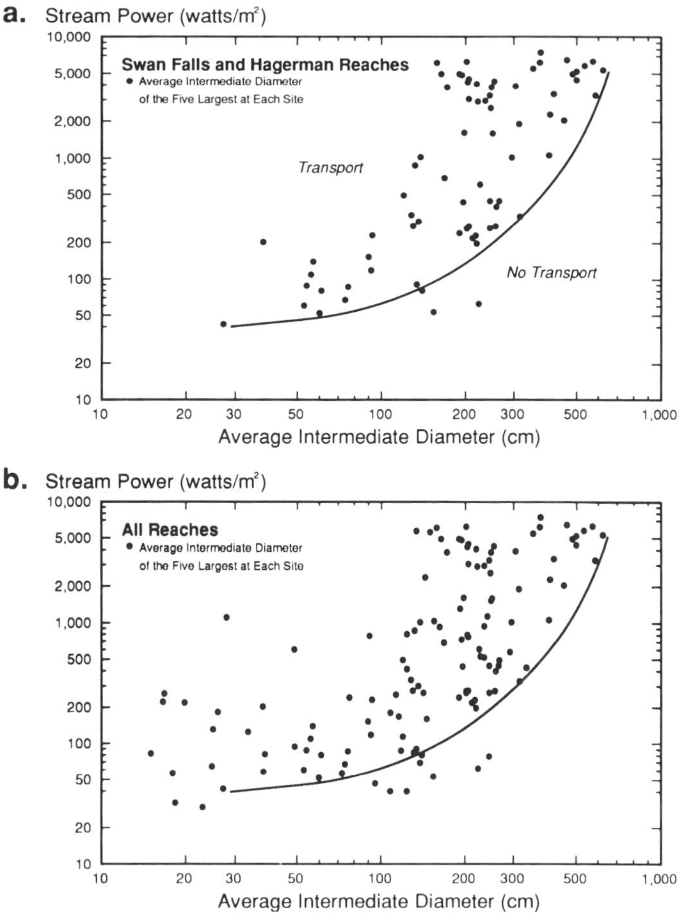

Figure 67. Maximum boulder size and local unit stream power magnitudes. Each point represents the average intermediate diameter of the five largest boulders (in terms of mass) at a measurement site. The local hydraulic conditions were derived from the step-backwater calculations at each measurement location as illustrated in Figure 53. The lower envelope defines the minimum hydraulic conditions for sustained transport for boulders deposited by the Bonneville Flood. The results from the Hagerman and Swan Falls reaches are probably more reliable because the step-backwater flow reconstructions in these reaches provide better matches to the geologic evidence of maximum flow stages. a, Swan Falls and Hagerman reaches. b, All reaches. The envelope curve shown here is the same as shown in a.

Costa, 1983). To aid in comparison, I have portrayed the results from the Swan Falls and Hagerman measurements in the same manner (Fig. 68). Boulder size has been compared to the hydraulic parameters of local flow depth (Fig. 68a), local flow velocity (Fig. 68b), and the derived parameters of shear stress (Fig. 68c) and unit stream power (Fig. 68d). The resulting relations for the Bonneville data must be viewed cautiously; the data were collected and organized with a goal of defining a lower limit of sustained transport, not average conditions associated with deposition. As stated previously, the boulder deposits probably do not all represent minimum transport conditions at peak discharge, and in many locations, boulder sizes representing maximum flow competence were probably absent. Nevertheless, there are correlations (high r^2 values) between boulder size and all of the hydraulic parameters except depth:

$$d = 22.8D_I^{0.13},\ r^2 = 0.04; \tag{6}$$

$$v = 0.29D_I^{0.60};\ r^2 = 0.52, \tag{7}$$

$$\tau = 0.33D_I^{1.12},\ r^2 = 0.50, \tag{8}$$

and,

$$\omega = 0.1D_I^{1.71};\ r^2 = 0.52, \tag{9}$$

where D_I is the average intermediate diameter (measured in centimeters) of the five largest boulders at a site, d is the flow depth (m), v is the flow velocity (m·sec^{-1}), τ is the shear stress (N·m^{-2}), and ω is the unit stream power (watts·m^{-2}). Also plotted are relationships for coarse particle transport determined by other workers over their applicable ranges (Baker and Ritter, 1975; Costa, 1983; Williams, 1983). Direct comparisons between the Bonneville Flood boulder measurements and the other analyses are not fully appropriate because of the different manners in which the data were obtained and treated, and the much greater flow depths and particle sizes in the relations derived from the Bonneville Flood data. Nevertheless, the results stimulate discussion regarding their differences.

The presented lines of Costa (1983) are regression lines *through* the data he compiled for velocity, unit stream power and shear stress to define critical entrainment thresholds (Figs. 68b–d). Costa stated that parallel lines *under* the data (not shown in Fig. 68) represent minimum entrainment values and are more appropriate for paleohydraulic analysis. For a more direct comparison with the Bonneville Flood boulder transport data, however, only the lines through Costa's data are shown. The lines of Williams (1983) are those drawn *under* his plots of data for observed clast motion and were proposed to represent minimum values of flow parameters associated with clast movement. For velocity, shear stress, and stream power, Williams's limiting lines plot in similar positions to Costa's (1983) minimum entrainment lines (not shown) because they are based, to a large extent, on the same data. Baker and Ritter (1975) restricted their analysis to shear-stress conditions and the regression line they determined from their compiled data is also shown in Figure 68c. There was also substantial overlap in the data used by Baker and Ritter (1975) and the studies of Costa and Williams; however, Baker and Ritter's analysis also contains measurements from the Missoula Floods (Baker, 1973).

For shear stress and unit stream power, the lines proposed by Williams, Baker and Ritter, and Costa (Fig. 68c, d) plot well above regression lines through the Bonneville Flood data: The regression lines of Costa (1983) and Baker and Ritter (1975) predict hydraulic conditions about 10 times more intense than the trend defined by the Bonneville Flood data, whereas the lower

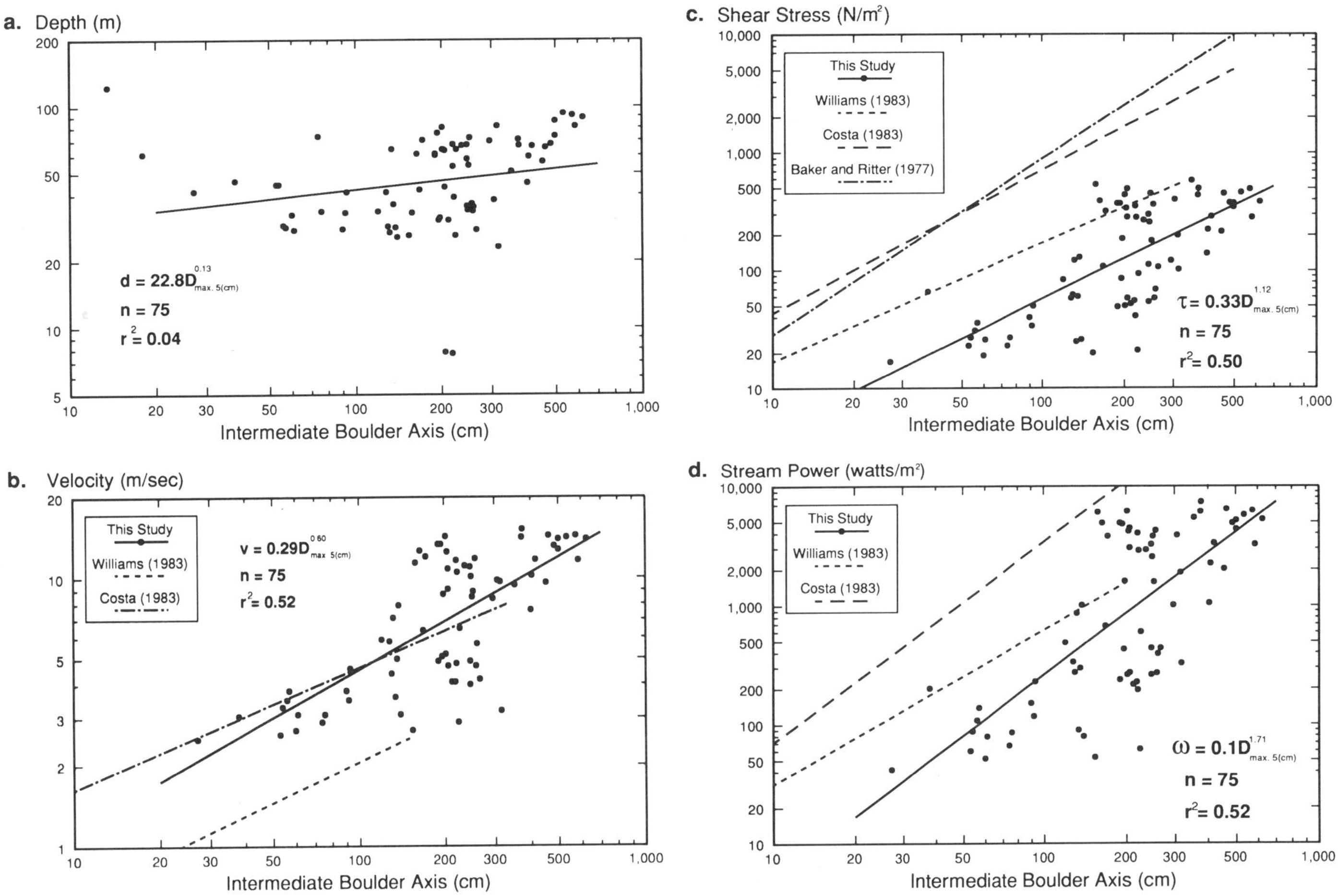

Figure 68. Power law relations between average intermediate diameter of the five largest boulders at a site for the Swan Falls and Hagerman reaches and local calculated hydraulic parameters. Relationships developed by other workers are also shown over their appropriate ranges. a. Depth. b. Velocity. c. Shear stress. d. Stream power.

limiting envelopes of Williams (1983) predict flow conditions two to three times stronger.

There are probably several reasons for the differences. Two factors stand out, however: (1) the difference between the flow strength necessary to initiate particle motion and that needed to sustain motion; and (2) the use of energy-slope values (instead of water-surface or channel slopes) in calculating the hydraulic variables at sites of boulder deposition for the Bonneville Flood. As stated previously, most of the data in the compilations of Baker and Ritter (1975), Costa (1983), and Williams (1983) were collected specifically to determine critical conditions associated with initial particle motion. (Nevertheless, the collected data do come from the entire range of initial motion, transport, and depositional conditions.) Critical entrainment conditions are undoubtedly more intense than those associated with maintaining particle motion. Hjulström (1935, p. 320) and Sundborg (1956, p. 181) reported critical entrainment velocities 1.5 to 2 times greater than velocities associated with deposition of pebble and cobble clasts. This factor could account for part of the difference between the Bonneville Flood data and the earlier compilations.

Perhaps the more important difference between the data sets, however, is the use of local energy-slope values for calculating shear stress and unit stream power at sites of measured Bonneville Flood boulders, as opposed to the more typical substitution of channel or water surface slope. For deposition to occur, flow competence must be decreasing, either temporally or spatially. In hydraulic terms, flow must be either nonuniform or unsteady. Along the route of the Bonneville Flood, it is apparent that boulders were deposited in reaches of reduced stream power and energy slope (Fig. 55); this spatial variability in the flow was a major factor controlling deposition. Furthermore, the energy slope varied substantially over very short distances; for example, within the Hagerman reach, the calculated energy slope spans two orders of magnitude. Substituting the channel slope or the water-surface slope for local energy slope (Fig. 12) implicitly assumes uniform flow conditions. The distribution of Bonneville

Flood tractive deposits is strong evidence that this assumption is invalid in reaches of deposition; consequently, this substitution probably leads to substantial error in calculating shear stress and stream power in environments of spatially varied flow. The closer match between the velocity conditions for clast movement determined from the compiled data of Williams (1983) and Costa (1983) and the Bonneville Flood data (Fig. 68b) accords with the energy slope factor being critical because the energy slope does not influence the calculated values of flow velocity as strongly as it does for shear stress and unit stream power (equations [4] and [5]).

Suspended load deposits

The suspended load consists of the entrained particles that are not in continual contact with the bed, but are supported by the flow energy and dispersive stresses (Bagnold, 1966). Material in suspension has been divided into two categories: the suspended load *per se* and the washload. Qualitatively, the suspended load can be thought of as those particles whose positions in the flow column are maintained by upward-directed momentum flux (Bagnold, 1966). The washload consists of the finest sediment and is considered to be completely incorporated within the flow structure, moving at the same velocity as the flow and at equal concentrations throughout the flow column. The quantity of washload appears to depend on geologic factors, such as sediment supply, rather than on flow competence (Einstein and others, 1940; Komar, 1979).

During the Bonneville Flood, both of these components of the suspended load were apparently present within the main thread of flow and are preserved in two different depositional facies in backwater areas. Eddy deposits are thought to consist of material derived from the suspended load and are found in tributary mouths, canyon wall reentrants, and other locations of flow separation. Slackwater sediment probably represents accumulations of the washload and are only found in wide, backflooded basins and tributary valleys where flow was ponded.

Eddy deposits. The suspended load accumulated at sites of markedly reduced velocity, most commonly in tributary mouths, downstream of projections into the flow, and in canyon wall reentrants (Baker, 1973, p. 36–39; Webster and others, 1982). These are areas of flow separation, where streamlines diverge from the main thread of flow in the channel and become involved in recirculation zones or eddies (Fig. 69). Consequently, these are also areas of substantially lower flow velocity, causing rapid deposition of a large portion of the suspended load. Sediment is continuously pumped into these recirculation zones as long as the main thread of flow retains the competency to maintain a suspended load. This is an important process in creating local sand accumulations in bedrock rivers (Schmidt, 1990). Because eddy deposits can be formed at all altitudes in the flow column, they have been critical to the hydraulic analysis of the Bonneville Flood by providing minimum estimates of maximum flood stage. Along the Bonneville Flood route, recognized eddy deposits are restricted to reaches that had high velocities and unit stream power magnitudes.

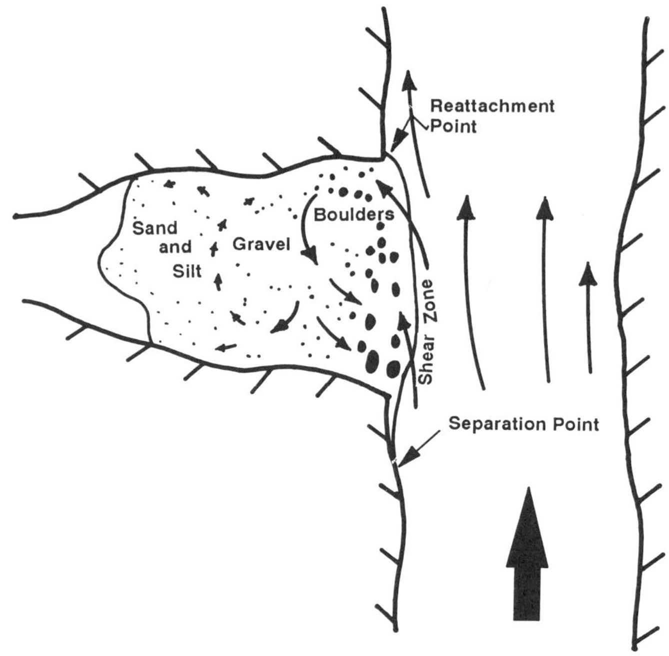

Figure 69. Schematic representation of flow conditions associated with the formation of eddy deposits. Length of flow vectors schematically represents relative flow velocities. Some of the terminology is from Schmidt (1990). The situation portrayed is thought to correspond to conditions associated with formation of the eddy bar across the river from the Initial Point bar (Fig. 37).

Form, stratigraphy, and depositional environment. Eddy deposits occur in a variety of settings depending on local geometric control of the flow recirculation pattern within a flow separation zone. However, a typical case is that of a deposit at the mouth of a tributary valley or canyon reentrant. One such deposit formed in the Swan Falls reach near the Initial Point bar (Fig. 37). At these locations, eddy deposits commonly consist of a ridge of relatively coarse sediment proceeding downstream from the point of flow detachment. In some cases, this ridge may completely block the tributary valley. Behind this ridge, up the tributary valley or within the canyon reentrant, there are generally finer and better-sorted deposits. The ridge of coarser sediment is probably near the eddy front, the shear zone between the main thread of downstream-directed flow and recirculating flow involved in the eddy (Fig. 68).

Another common form of eddy deposits left by the Bonneville Flood, especially in Hells Canyon, are gravel and sand accumulations on the upstream sides of small divide crossings marginal to the main channel. There are several examples of such deposits at Pittsburg Landing (Fig. 43) and near Dug Bar (Fig. 44). At these locations, the tops of the eddy deposits are generally graded to the bottom of the divide.

The best exposures of eddy deposits are within the Snake

River Plain where they have been extensively exploited for gravel. At these locations, they are composed of well-sorted basaltic sand and gravel deposited in chaotically oriented festoon crossbeds (Fig. 70). The large crossbeds and the variety of bed orientations are probably the result of changing flow patterns within the recirculation zone as the deposit grew and floodstage changed. The sizes of clasts in these deposits probably depended on flow competence in the main channel (as is discussed later). In Hells Canyon, the eddy deposits primarily consist of angular gravel derived from local sources. Exposures are rare; however, observations at isolated locations indicate that they are poorly sorted and are deposited in inclined beds dipping away from the main channel.

Slackwater deposits. In wide basins and other ponded areas along the flood route, there are areally extensive deposits of laminated clay, silt, and sand (Fig. 71) (Malde, 1968, p. 19; Webster and others, 1982; Othberg and Fosberg, 1989; W. E. Scott, 1989, personal communication). Similar deposits in the Channeled Scabland continue to be the subject of controversy regarding interpretation of depositional processes and the frequency of Missoula Flooding. (see Baker and Bunker [1985] and Waitt [1985] for recent reveiws.) Along the Bonneville Flood route they are simply interpreted to have been deposited in areas of flow stagnation or quiescence and to represent the washload component of the suspended load. In some locations, however, they seem to grade (texturally and spatially) to eddy deposits near the main thread of flow (e.g., near the mouth of the Grande Ronde River Valley; Fig. 47). Consequently, the distinction is not always clear.

Form, stratigraphy, and depositional environment. Slackwater deposits are found in areas that were inundated by standing or slow-moving water. These areas include the northern margin of the Burley Basin (Fig. 19), the King Hill Basin (Malde, 1968, p. 19), some of the flooded basins between the Hagerman and Swan Falls reaches, the backflooded river valleys between the Swan Falls and Farewell Bend (Othberg and Fosberg, 1989), and in backflooded valleys near Lewiston. The largest accumulations are between Swan Falls and Farewell Bend where the Boise, Payette, and Owyhee River valleys are mantled by fine-grained laminated sediment up to altitudes of 744 m (2,440 ft) (Othberg and Fosberg, 1989; K. Othberg, 1991, personal communication).

The distribution of slackwater sediment is restricted to areas of ponded water; however, an upstream source of washload also appears to be an important factor for their presence. Most of the sediment in the deposits appears to be reworked loess, although there are sandier layers that contain basalt grains. Loess is ubiquitous on the Snake River Plain with thicknesses exceeding several meters at many locations, especially on the eastern part of the Plain (Lewis and Fosberg, 1982; Pierce and others, 1982). This loess contains particle sizes ranging from fine silt to very fine sand (0.008 to 0.125 mm), with coarse silt (0.031 to 0.062 mm) constituting more than 50% of the distribution (Lewis and Fosberg, 1982). A substantial amount of loess was probably entrained by the flood over the broad basaltic uplands north of Massacre Rocks and along the route of Eden channel (Figs. 3, 21). In these reaches, velocities were high and the flow was over wide swaths of thick loess deposits that mantle the basalt. Consequently, slackwater sediment was deposited primarily in basins

Figure 70. Photograph of an exposure of an eddy deposit within the Hagerman reach (Fig. 28). Coarse sand to 3- to 4-cm gravel was deposited in festoon crossbeds. Larger clasts of less dense rock types are also present.

Figure 71. Exposure of slackwater deposits near the Boise River (Fig. 3). The deposits consist of laminated silt and very fine sand that was probably derived from loess entrained in the flow as washload.

downstream from these source areas. The hydraulically ponded valley upstream of the Portneuf Narrows contains few slackwater deposits because the flow in this reach probably contained little washload.

The most typical stratigraphic style of Bonneville Flood slackwater sediment are thin- to several-meter thick beds of clay, silt, and sand passively deposited over the prior ground surface, generally loess. Bedding and particle size vary: the finer fraction (clay- and silt-sized particles) generally exhibit planar laminae and weather to a blocky structure developed along sandy parting planes (Fig. 71). The coarse silt and sand-sized sediment commonly show ripple-drift cross-lamination, indicating rapid deposition (Jopling and Walker, 1968). Deposits closer to the Snake River are generally coarser and contain sedimentary structures indicative of higher flow velocities than do more distal deposits. The presence of ripple structures indicates that some of the sediment was tractively transported immediately prior to deposition; therefore, in these places, the deposits are not truly "slackwater" sediment, but were moved by local currents to depositional sites. Nevertheless, this material was probably part of the suspended load while in the main thread of the flow.

A distinct type of low-energy deposit are "rhythmites" (Figs. 72–74). These are multiple sequences of graded silt and sand (and locally, gravel) deposited in a distinctly repetitive fashion. Each sequence grades upward from plane-laminated coarse sand to ripple-laminated sand and silt. Within many of the ripple-laminated sequences, ripple orientations indicate changing (at least 90°) flow directions. Tops of many of the beds have flame structures that protrude into the overlying sequence (Fig. 72). The lower boundaries are generally conformable, although locally there is evidence of scour into the underlying sequence. Rhythmites near Lewiston show cut-and-fill relations with the fill sequences consisting of several of thinner units (Fig. 74).

There are two areas where rhythmites are exposed; near the town of Hammett, downstream of the Hagerman Reach (Figs. 28, 72, 73), and near Lewiston (Figs. 48, 74). The rhythmites near Lewiston have been studied previously because of their probable relationship to the Missoula Floods (Webster and others, 1982; Waitt, 1985). The Hammett rhythmites (Figs. 72, 73) are coarser and thinner than the Lewiston-area rhythmites (Fig. 74); however, the general sedimentologic pattern is similar.

The rhythmites in the Lewiston area have been interpreted (Waitt, 1980, 1985; Webster and others, 1982) to be equivalent to the Touchet Beds in eastern Washington (Bretz, 1929; Baker, 1973, p. 42–47) and are inferred to have been deposited by backflooding up the Snake River valley during one or more of the great Missoula Floods (Webster and others, 1982) (Fig. 1). Waitt (1980, 1985) noted evidence for subaerial exposure between some of the rhythmites and concluded that each individual rhythmite was left by a separate Missoula Flood. At Tammany Bar (Figs. 48, 74), the Bonneville Flood point bar is capped by 17 to 21 rhythmites (Webster and others, 1982; Waitt, 1985), perhaps resulting from a like number of Missoula Floods inundating Tammany Bar (280 to 300 m; 920 to 980 ft) subsequent to the Bonneville Flood. Recent stratigraphic and sedimentologic study of rhythmites in the Channeled Scabland indicates that many were indeed the result of individual floods, but some were probably the result of pulses within floods (Smith, 1992). The rhyth-

Figure 72. A sequence of graded beds deposited by the Bonneville Flood near Hammett (at the downstream end of the Hagerman reach [Fig. 28]). The graded beds at Hammett are thinner than those deposited by Missoula Floods near Lewiston; however, they show similar, repeated grain-size variations and sedimentary structures. Note rock hammer for scale.

Figure 73. Photograph of a single graded bed (shown by the arrows) within the Hammett deposit (Fig. 72) showing the typical stratigraphic sequence of plane-laminated coarse sand grading up to trough-crossbedded fine sand and silt.

mites near Hammett (Figs. 72, 73) were undoubtedly deposited by the Bonneville Flood, indicating that a single flood can emplace rhythmites.

Hydraulic conditions associated with the suspended load. The hydraulic conditions associated with depositing the suspended load and slackwater deposits cannot be directly estimated from the step-backwater calculations because, in defining the channel geometry for the hydraulic flow reconstructions, areas of flow stagnation and separation were purposely excluded. Nevertheless, analyses by other workers of flow conditions capable of entraining clasts into suspended load and for forming the sedimentary structures preserved in the slackwater deposits provide for indirect assessments of local conditions for suspended-load entrainment and deposition associated with the Bonneville Flood.

Bagnold (1966) proposed that the suspended sediment load (including the washload) in a flow is produced by local elements of net upward-directed momentum flux within the flow that opposes the gravitational tendency for particles to settle. Bagnold (1966) expressed this balance, rearranged by Komar (1980), as:

$$\frac{w_S}{u_*} = k, \qquad (10)$$

where w_S is the settling velocity for a particular grain size, u_* is the shear velocity, and k is a constant that discriminates between bedload and suspended load transport modes for a particle of settling velocity w_S. The shear velocity is thought to be representative of upward-directed velocity components of turbulent eddies and is a function of boundary shear stress:

$$u_* = \sqrt{\left(\frac{\tau}{\rho}\right)}, \qquad (11)$$

where ρ is the density of the fluid and τ is the shear stress as defined in equation 4. Combining equations 10 and 11 yields:

Figure 74. Photograph of rhythmites on top of Bonneville Flood tractive sediment at Tammany bar near Lewiston (Fig. 48). The foresets in the Bonneville Flood deposits dip up the valley of Tammany Creek, indicating flow up the Tammany Creek valley as the bar was formed. The rhythmite section is about 10 m thick and is composed of 20 to 21 rhythmites (Waitt, 1985). Each rhythmite has been interpreted to represent an episode of backflooding up the Snake River from large Missoula Floods crossing the Channeled Scabland of eastern Washington (Fig. 1) (Waitt, 1980, 1985; Webster and others, 1982). The rhythmites are capped by about 1 m of postflood loess.

$$\tau_c = \frac{\rho w_s^2}{k^2}\,; \tag{12}$$

where τ_c is the critical shear-stress value for initiating suspension. Bagnold (1966) and Komar (1980) proposed a k value of 1.25 for the threshold for entrainment of clasts into suspension. Furthermore, Komar (1980) suggested a k value in the range of 0.05 to 0.10 for discriminating particle sizes incorporated into the washload. A separate threshold for the washload based on flow energy is counter to the conceptual notion that the washload is more related to geologic setting than to flow competence; however, these threshold values are consistent with washload concentrations in measured flows (Komar, 1980). With these limits, the conditions of the Bonneville Flood can be evaluated (Fig. 74), realizing, of course, that extrapolation is required.

Maximum shear stresses achieved by the Bonneville Flood were on the order of 2,500 N·m^{-2} (Table 3). In the steeper reaches, the flow reconstructions indicate sustained values of over 1,000 N·m^{-2} for several kilometers. According to the relations discussed above and shown in Figure 75, these shear stresses would have entrained clasts with diameters as long as 10 to 20 cm into suspended load. Washload would have been restricted to sand-sized material (less than 1.5 to 2.0 mm). These results are consistent with the particle sizes incorporated in the eddy deposits and the slackwater sediment.

Local conditions at some sites of slackwater sediment deposition can be estimated from studies relating current ripple formation to local hydraulic conditions. Middleton and Southard (1984, p. 280–295) reported that ripples in sand form under average flow velocities of 0.2 to 1.0 m·sec^{-1}. Allen (1968, p. 144) suggested corresponding unit stream power values of 0.1 to 1 watts·m^{-2}. These values seem reasonable for the ponded areas of the Bonneville Flood and are indicative of the extremely low energy levels that existed for parts of the flood route. Considering the complete spectrum of Bonneville Flood deposits, unit stream power values spanned about five orders of magnitude in depositing clasts with diameters ranging from 10^{-4} to 10^3 cm.

Erosional features and processes

Some of the most impressive landscapes on the Snake River Plain are the scabland tracts eroded by the Bonneville Flood. Gilbert (1890, p. 176) drew his analogy between the "Bonneville river" and the Niagara River on the basis of similar erosion features carved into the basalt in Marsh Creek valley (Fig. 17). Malde (1968, p. 20) provided a general description of "scabland" features on the Snake River Plain, and in particular, described (p. 25–30) the spectacularly eroded terrain near the city of Twin Falls where extensive erosion into the basalt flows has produced

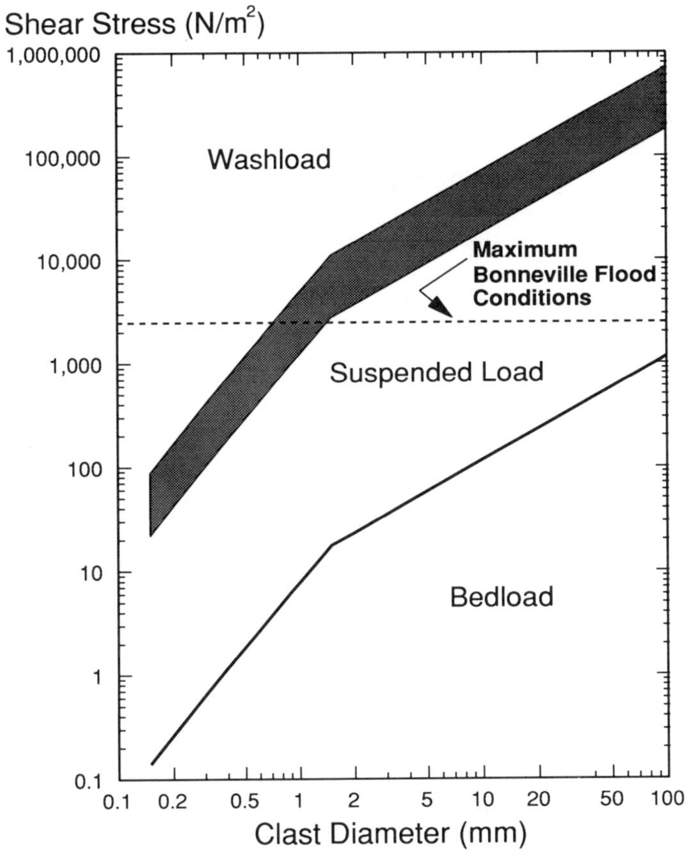

Figure 75. Local shear stress values required for entrainment of suspended load as given by equation 12 for the relationship between particle diameter and fall velocity (w_S) provided by Raudkivi (1976, p. 10). Maximum Bonneville Flood shear stresses are predicted to have entrained washload particles as large as about 1 mm and suspended clasts of 10 to 20 cm. These values are consistent with the maximum particle sizes observed in the slackwater deposits (Fig. 71) and the eddy deposits (Fig. 70). Adapted from Komar (1980).

large cataracts (Fig. 11) and butte-and-basin topography. These types of erosional features seem to be best preserved and recognized on basalt. Undoubtedly other rock types, especially those in Hells Canyon, were also extensively eroded; however, definitive evidence of fluvial erosion was more difficult to recognize in nonbasaltic rocks.

Scabland topography. Bretz (1923) borrowed the term "scabland" from local settlers in eastern Washington to describe the craggy and eroded basalt flows scoured by the Missoula Floods. Very similar landscapes formed along the Bonneville Flood route.[6] I have classified "scabland" into two categories. (1) The most abundant along the Bonneville Flood route are areas of intact surfaces of basalt flows that have had their loessal cover removed. These *stripped basalt* surfaces are covered by a thin (less than about 1 m) veneer of postflood loess, and commonly there are scattered boulders and small gravel accumulations in areas of local deposition. For example, large tracts within the Eden channel were stripped several meters of loess (Fig. 26). (2) Indicative of more intense erosion are areas of *butte-and-basin scabland* topography, where portions of the uppermost basalt flow(s) have been completely removed. The remaining topography is of knobs or mesas of basalt surrounded by closed depressions (e.g., near the Rock Creek constriction [Fig. 31]). The East of Eden potholes (Figs. 26, 27) are an extreme case of butte-and-basin scabland. Within the Portneuf Narrows, basalt flows following the valley bottom were completely eroded away (Fig. 7). The most intense erosion was probably associated with development of the cataract complexes. These are large channels incised into the basalt that head in (multiple) horseshoe-shaped headcuts (Fig. 11) (Baker and Komar, 1987). Several cataracts formed where flow from the Eden channel spilled into the Snake River canyon between physiographic Twin Falls and the upstream end of the Hagerman reach. There are also cataracts near Massacre Rocks and near Crane Falls (Malde, 1968, p. 37–40).

Erosion processes. Processes that erode bedrock are not well understood because they generally occur at rates that preclude direct observation: either operating over extremely long time periods at very slow rates, or over extremely short periods at very high rates. Even if one could witness bedrock erosion, the magnitude of the forces involved defies direct measurement. Consequently, there is little quantitative information of erosional processes and thresholds in natural bedrock fluvial systems. However, erosional features along the Bonneville Flood route can, in certain places, be related to local hydraulic conditions calculated from the step-backwater flow reconstructions.

Baker and Komar (1987) qualitatively described the development of channeled scabland as a five-phase process involving different erosional mechanisms (Fig. 76). The sequence of erosional processes postulated by Baker and Komar (1987) are consistent with observations made in flume experiments with simulated bedrock fluvial systems (Shepherd and Schumm, 1974). Erosional mechanisms thought to be important in scabland development include longitudinal roller vortices and "kolks." Kolks (Matthes, 1947; Baker, 1984) are vertical vortices, like dust devils, and are thought to be capable of extensive erosion.

Cavitation. Another process that may be important in bedrock fluvial systems and was probably a major erosional agent in both the Bonneville and Missoula Floods is cavitation (Barnes, 1956; Baker, 1984; Baker and Costa, 1987; Wohl, 1992b). Cavitation as a geologic process has not received much attention; most of the available information comes from the engineering literature and its operation in natural systems is not well understood. However, cavitation has been documented as causing severe erosion of dam spillways and has been postulated to initiate the formation of

[6]In building a case for a nonflood origin of the Cheney-Palouse scabland tract in the Channeled Scabland, Flint (1938, p. 492) described the scabland topography flanking Shoshone Falls, noting the similarity to features in eastern Washington; ". . . yet I am not aware that unusual floods have been held to have affected the upper Snake River."

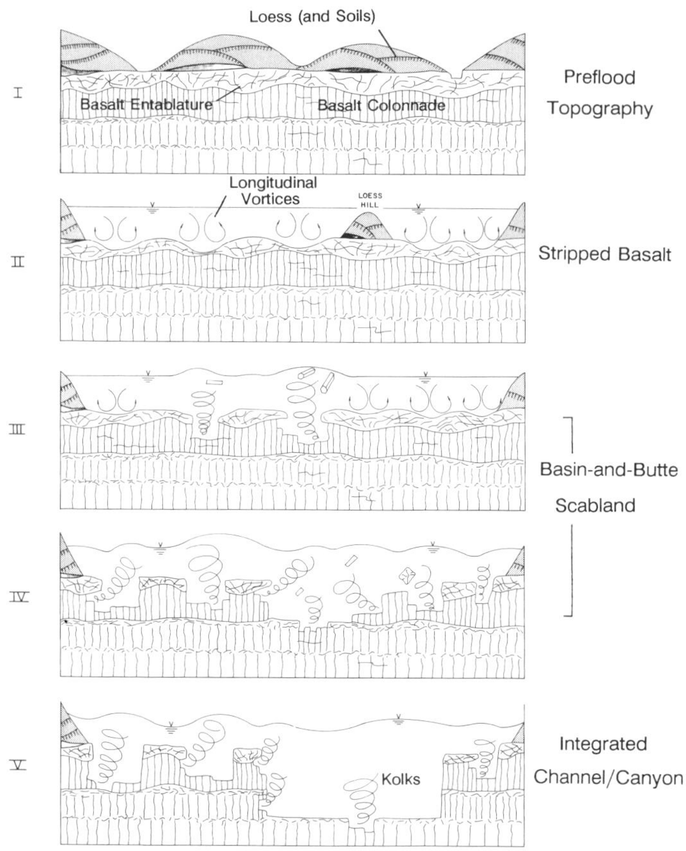

Figure 76. Hypothetical sequence of erosion for the development of scabland topography. Phase I represents initial preflood topography of loess-covered basalt flows. Phase II is the removal of the loessal cover leaving a stripped basalt surface (Fig. 23 shows an example of stripped basalt surfaces). Phases III and IV represent the development of butte-and-basin topography, perhaps in association with kolks and cavitation. The butte-and-basin topography near the Rock Creek constriction (Fig. 31) is an example of this stage of erosion. Phase V is the creation of an integrated channel. Along the route of the Bonneville Flood, flood-formed channels are generally restricted to cataracts that migrated headward from the canyon rims (for example, the Blue Lakes cataract complex shown in Fig. 11). The East of Eden potholes (Figs. 26, 27), however, may be one site of incipient canyon formation resulting from local scour. Modified from Baker and Komar (1987).

potholes in bedrock channels (Barnes, 1956). Cavitation bubbles (of water vapor) form where the local fluid pressure becomes less than the vapor pressure of the fluid. Areas of reduced fluid pressure can be caused by local velocity perturbations; as the fluid accelerates around roughness elements, local fluid pressure decreases correspondingly. In deep and fast flows where there are local velocity perturbations, the local fluid pressure can drop to the vapor pressure at locations of flow acceleration, allowing formation of cavitation bubbles. When these bubbles move to zones of greater fluid pressure (decreased velocity or increased depth), they become unstable and implode. The attendant forces have not been measured in natural systems; however, local pressures exceeding 3,000 MPa have been measured in small-scale experiments (Barnes, 1956).

Approximate conditions in which cavitation will occur can be determined by considering two fluid elements in a flow column, one at the top of the flow and the other near the bottom (the following discussion is partly taken from Barnes [1956]). If the flow is hydrostatic (not a completely appropriate assumption for this analysis, but the results should be of the correct magnitude), the mechanical energy of the two fluid elements is equal and the conditions for which cavitation will occur can be described by a form of the Bernoulli equation:

$$\frac{v_1^2}{2g} + \frac{p_1}{\gamma} + d = \frac{v_2^2}{2g} + \frac{p_2}{\gamma}, \tag{13}$$

where v_1 is the velocity of a fluid element at the top of the flow (assumed to equal the mean flow velocity), d is the flow depth, p_1 is the atmospheric pressure, v_2 is the local, accelerated, velocity at the point of cavitation (assumed to be near the bottom of the flow), and p_2 is the fluid vapor pressure. Equation 13 prescribes the conditions at the bottom of the flow where the local fluid pressure equals the vapor pressure. For hemispherical projections into the flow, local velocities can equal 1.5 times the mean velocity (in terms of equation 14, $v_2 = 1.5\ v_1$) (Rouse, 1946, p. 32). Other, less fluid dynamic shapes may cause local velocities as great as 2.4 times the mean velocity (Barnes, 1956). Solution to equation 13 takes the form

$$v_c = a\sqrt{b + d}\,, \tag{14}$$

where a is inversely related to the magnitude of local velocity perturbations; b is a function of environmental factors such as atmospheric pressure, fluid density, and temperature; d is flow depth; and v_c is the threshold mean flow velocity above which cavitation can be expected to occur.

For local velocity accelerations 1.5 and 2.4 times the mean flow velocity, the resultant a values are 4.0 and 2.0, respectively, which may approximate the range of conditions for flows in natural channels. For extremely smooth surfaces, where local velocity accelerations are small, a quickly becomes very large, requiring very large mean flow velocities for critical cavitation conditions to occur. Postulated environmental conditions for the Bonneville Flood on the Snake River Plain (an altitude of 1,000 m, fluid temperature of 4.4°, and clear water) result in a b value of 8.4. The corresponding solutions are shown in Figure 77.

As shown in Figure 77, flow velocity and depth combinations for which cavitation is likely can be separated from those combinations where cavitation probably does not occur. The area between these two fields, bounded by conditions for a values of 2.0 and 4.0, represent depth-velocity combinations where the potential for cavitation depends on the degree of local velocity accelerations. This, in turn, depends primarily on the surface roughness; for a given mean flow velocity and depth, the rougher the surface, the greater the magnitude of local velocity accelerations, and the more likely it is that cavitation will occur.

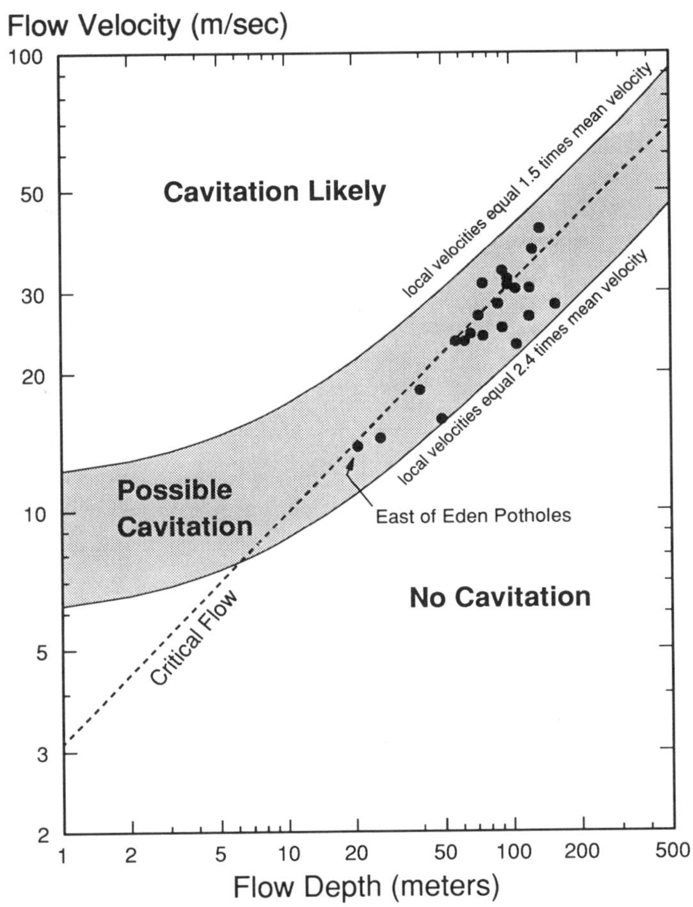

Figure 77. Approximate hydraulic conditions required for potential cavitation on the Snake River Plain. Maximum local bottom velocity accelerations are presumed to result in instantaneous velocities of 1.5 to 2.4 times the mean flow velocity, depending on local roughness. Depth-velocity combinations for some of the major Bonneville Flood constrictions (Table 3) plot well into the field of potential cavitation, as does the site of the East of Eden potholes. The step-backwater modeling results are constrained by critical flow conditions; the actual conditions at these constrictions would plot higher into the field of potential cavitation if the flow was supercritical.

The potential for cavitation is also dependent on flow depth; at flow depths less than 10 m, cavitation occurs only at flow velocities greater than 6 to 15 m·sec^{-1}, depending on the magnitude of the local velocity accelerations. For flow depths greater than 20 m, the critical cavitation velocity is a power function of depth. Consequently, cavitation is a process that is scale-dependent. Indeed, cavitation is more likely in deep flows because of the close relationship between critical cavitation velocity and critical flow conditions. At flow depths greater than 20 m, the zone of critical cavitation velocities brackets flow conditions of minimum specific energy (critical flow) (Fig. 78). For most bedrock stream systems, and probably for most of the Bonneville Flood route, critical flow is a limiting condition for maintaining stable hydraulic conditions. In spatially varied channel geometries where there are resistant boundaries, supercritical flow cannot be maintained for long distances because locally narrow segments hydraulically impound flow to subcritical depths. Therefore, for bedrock stream systems, cavitation is much more likely to occur where flow depths exceed 10 to 20 m. At flow depths less than 10 m, flow conditions well into the supercritical range are required for cavitation to occur.

Erosion by cavitation is likely to be a progressive process. For locations where erosion by cavitation increases local roughness, as probably is case on basalt surfaces, continued cavitation is enhanced, once started, because of the increased magnitude of local velocity accelerations. Furthermore, sites of higher fluid pressure, where the cavitation bubbles implode, are likely to be the slightly deeper areas. These areas enlarge and deepen as the cavitation implosions fracture, loosen, and remove pieces of the channel bottom. As they deepen, the enlarging depressions become increasingly subject to cavitation implosions because of increasing local fluid pressure associated with greater flow depths. I envision this process continuing until the scour holes enlarge and become interconnected to the extent that the local cross-section conveyance increases and the mean flow velocity drops below that necessary to maintain cavitation.

Along the Bonneville Flood route, cavitation probably occurred at all locations that experienced critical flow. Some of the more intense depth-velocity combinations are shown in Figure 77. It is difficult, however, to identify erosional features that might be uniquely attributable to cavitation. One exception could be the East of Eden potholes (Figs. 26, 27). The East of Eden potholes consist of several pits or depressions, as much as 40 m deep, excavated into the upper basalt flows. Some of them are circular in plan, but some are elongate and oriented parallel to flow (Fig. 26). These scoured pits may represent an early stage of canyon formation; if flow in the Eden channel had continued longer, these potholes may have coalesced into an integrated canyon or inner channel, as illustrated in phase V of the model of Baker and Komar (1987) (Fig. 76).

Moreover, the local hydraulic conditions, as determined from the step-backwater calculations, are consistent with cavitation being an agent in their formation. Because of the comparatively low flow depth and velocity in the Eden channel, maximum calculated unit stream power values (23,600 watts·m^{-2}) and shear stress (1,700 N · m^{-2}) were not exceptional with respect to the other constrictions (Table 3). However, because of the high flow velocities, the site of the East of Eden potholes plots well into the field of potential cavitation (Fig. 77), and, consequently, very local forces may have been an order of magnitude or more greater. The canyon bottom along the Murtaugh reach, the youngest and steepest segment of the Snake River Canyon, also has abundant potholes of all different sizes. Some portions of the low flow channel appear to follow routes of large, interconnected, depressions (Fig. 22) that may have been formed by cavitation during the Bonneville Flood.

Unit stream power values associated with erosion. Cavitation is one erosional process that can be evaluated with the step-backwater calculations. However, empirical observations can provide a feel for the general magnitude of stream power values associated with erosional terrains. Site-specific stream power magnitudes at sites of erosion, compared with values from sites of deposition, are shown in Figure 78. Local values of unit stream power are calculated for 135 sites of butte-and-basin scabland topography in a manner similar to that for depositional sites (Fig. 53). Stream power values at sites of deposition are directly from the analysis of the hydraulic conditions at boulder measurement sites.

As expected, in comparison with depositional areas, the unit stream power values at sites of identifiable erosion (butte-and-basin scabland and eroded basalt flows) are displaced toward greater values, although there is substantial overlap (Fig. 78). Local stream power values at sites of erosion range from 300 watts·m^{-2} to 30,000 watts·m^{-2}. Undoubtedly, erosion occurred at higher stream power values as well. In contrast, almost all sites of deposition had stream power values less than 6,000 watts·m^{-2}.

More striking and perhaps more relevant is the overall correspondence of depositional and erosional sites with reach-scale variations in unit stream power (Fig. 55). Locations of butte-and-basin scabland topography, and areas of complete erosion of basalt flow units, are within reaches that had high unit stream power values, generally in excess of 20,000 watts·m^{-2}. This might be a more representative threshold for the energy expenditure required to erode the competent basalt of the Snake River Plain.

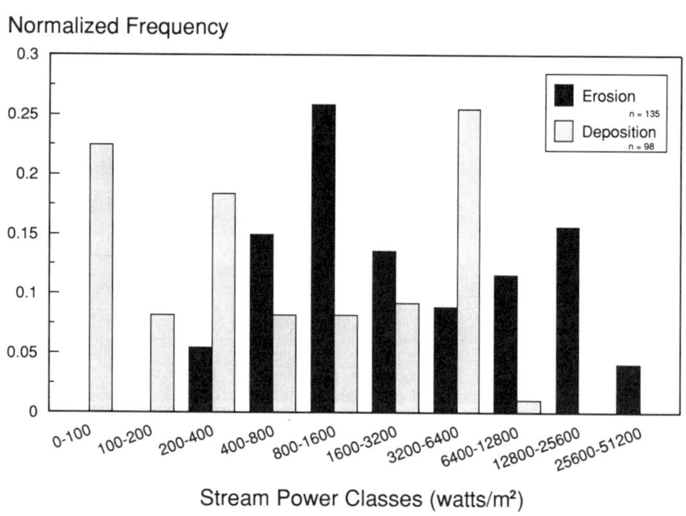

Figure 78. Site-specific stream power values at sites of butte-and-basin topography and tractive deposit deposition. The depositional sites are those used in the analysis of boulder deposition and are restricted to Snake River Plain locations. The stream power values at sites of erosion should be viewed cautiously; these were probably sites of evolving channel geometry and the reconstructed hydraulics, associated with the present topography, may not reflect conditions during erosion. Nevertheless, it is apparent the sites of erosion were generally at locations of higher stream power values than depositional sites.

SUMMARY AND CONCLUSIONS

The setting of the Bonneville Flood provides an excellent field laboratory to investigate an extreme geomorphic event. The extensive physical record of the flood provides firm grounding for hypotheses of flood processes. Step-backwater reconstruction of flow conditions furnishes a means to evaluate some of these features and processes quantitatively. Step-backwater calculations are a simplified deductive treatment of open-channel flow; like many tractable deductive treatments of geomorphic phenomena, such methods only provide relevant results under certain circumstances. The abundant geologic evidence of maximum stages along the Bonneville Flood route helps define the limits of applicability for step-backwater calculations in a real geomorphic system. Working within these limits, some insights can be made regarding the hydrology, hydraulics, and geomorphology of the Bonneville Flood.

The most reliable discharge estimates along the flood route are from reaches that had constricted channel segments of critical flow and where there is good evidence of maximum flood stages in the hydraulically impounded reaches upstream. The peak discharge of the flood at the outlet at Red Rock Pass was 1.0 ± 0.15 million m^3·sec^{-1}. At the downstream end of the study reach, the peak discharge attenuated to 0.57 to 0.62 million m^3·sec^{-1}. Much of the attenuation was probably the result of flood-water storage in the backflooded valleys between Swan Falls and Farewell Bend (Fig. 3).

The high-water evidence and the flow reconstructions provide evidence for the profound spatial variation in local hydraulic conditions and energy expenditure along the flood route. Maximum channel stream power values ranged from less than 10 watts·m^{-2} to more than 100,000 watts·m^{-2}. In the modeled reaches, about 50% of the flood energy at peak discharge was expended over 6% of the distance. Moreover, these spatial variations in energy expenditure clearly relate to the distribution of flood features. Erosion of basalt generally corresponded to reaches with maximum stream power magnitudes exceeding 20,000 watts·m^{-2}. Bedload was deposited in reaches of declining and low unit stream power values (less than 6,000 watts·m^{-2}).

In appropriate settings, the step-backwater calculations provide estimates of site-specific hydraulic conditions that can be used for quantitative evaluation of some flood processes. Two examples from the analysis of the Bonneville Flood are (1) the evaluation of boulder deposition, and (2) the potential role of cavitation as an important erosional process. In contrast to most previous analyses of flow competence, the conditions associated with boulder *deposition* have been distinguished from the hydraulic conditions and processes responsible for *initiation of motion.* The calculated conditions associated with sites of boulder deposition define a limit of sustained transport that relates to particle size. Tractively transported boulders were deposited in reaches of declining flow strength resulting from spatial (and probably in some locations, temporal) changes in channel geometry.

Analysis of cavitation as an erosional process has generally been restricted to the realm of engineering and manmade structures. However, the hydraulic conditions attained by deep flows such as the Bonneville Flood are consistent with cavitation being an effective erosional process in highly energetic natural environments. Furthermore, some of the erosional features left by the Bonneville Flood, such as the East of Eden potholes, may be the geomorphic consequence of cavitation. Cavitation may indeed be an important process of bedrock erosion in certain environments, especially for deep flows near critical depth.

In summary, the sedimentology and geomorphology of the Bonneville Flood have similar elements to that resulting from floods of a wide range of scales in many types of fluvial systems. The eroded scabland landscapes and the large scale of the flood deposits are like those left by the great Missoula Floods which may have been the greatest Quaternary floods on earth. The bar forms and clast orientations are similar to those observed for pebbles and cobbles in flumes and small fluvial systems. The pattern of extreme spatial variations in the hydraulics is probably similar, albeit at a greater scale, to those in all natural resistant-boundary fluvial systems.

ACKNOWLEDGMENTS

This project was the focus of my doctoral research at the University of Arizona and represents the culmination of several years of graduate study. Many people have provided support, encouragement, and inspiration during my graduate student career, and my successful completion is due in large measure to their help. My dissertation committee members—Simon Ince, Robert Clark, Owen Davis, William Bull, and Victor Baker—have all contributed to this project. Vic Baker and Bill Bill provided advice and inspiration during my entire tenure in Tucson and have been major forces in the development of my geologic career.

Many fellow graduate students at the University of Arizona have helped with this project both in the field and at home. Sara Rathburn, Lisa Ely, Kirk Vincent, and Julie Woodward all spent time in Idaho and participated enthusiastically in measuring countless boulders. Vincent, as a native Idahoan, contributed the indigenous reasoning abilities and his homespun wisdom to many aspects of the project. Ely, as a counterpart and coinvestigator for a multitude of adventures, was a unique influence on this project's final form.

Several people and agencies in Idaho made my work along the Snake River more efficient and enjoyable. Art Seamons and his crew at the U.S. Forest Service in Hells Canyon National Recreation Area provided logistical support in an area of difficult access. Tracy Vallier of the U.S. Geological Survey provided support and information for understanding the geology of Hells Canyon. Willie Scott, also of the U.S. Geological Survey, shared his observations of the geology of Lake Bonneville and the eastern Snake River Plain. Kurt Othberg, Bill Bonnichsen, and Margi Jenks of the Idaho Geological Survey all directed me to important locations in southwestern Idaho. The Idaho Department of Fish and Game provided housing while I was working near Hagerman. Paul Link and Thomas Ore of Idaho State University provided geologic information, housing, and a spectacular aerial view of much of the flood route while I was working near Pocatello. U.S. Bureau of Land Management offices in Pocatello and Burley provided access to their aerial photo collections.

This project was inspired in large measure by Hal Malde's 1968 Professional Paper on the Bonneville Flood. His encouragement as well as a week in the field provided impetus and direction.

This project was funded in part by Sigma Xi—the Scientific Research Society; the Quaternary Geology and Geomorphology Division's Hoover Mackin Award), and the GSA's Penrose Research Grants; and by the National Science Foundation (Grant EAR-8805321). A National Research Council Postdoctoral Associateship with the U.S. Geological Survey, under the supervision of John Costa, has allowed me to put this Special Paper into final form.

Karen Demsey, my wife, participated in all aspects of this project, from start to finish, in the field and in the office. This project never would have been completed without her.

Reviews by Robert Jarrett, Ellen Wohl, Jasper Hardison, Richard Hoppin, Carol A. Thompson, Rene Munoz, and an anonymous GSA reviewer all substantially improved the manuscript.

APPENDIX 1. LIST OF SYMBOLS

Symbol	Quantity	Location Introduced
a	Coefficient related to magnitude of local velocity accelerations for determining the potential for cavitation	Equation 14
b	Coefficient related to local environmental conditions for determining the potential for cavitation	Equation 14
C	Energy loss coefficient for estimating energy losses associated with channel expansions and channel contractions	Equation 2
d	Flow depth	Equation 1; Figure 12
D	Clast axis length	Figure 58
D_I	Average intermediate axis length for the five largest clasts at a measurement location	Equations 6-9
g	Acceleration of gravity	Equation 1; Figure 12
h_e	Head loss between a pair of cross sections	Equation 1; Figure 12
k	A constant that discriminates modes of sediment transport	Equation 10
L	Reach length	Equation 2; Figure 12
n	Mannings coefficient of roughness	Equation 3
p	Atmospheric pressure	Equation 13
R	Hydraulic radius	Equation 3
S_e	Energy slope (total mechanical energy loss)	Figure 12
S_f	Friction slope (energy loss due to flow boundary resistance)	Equation 2
u_*	Shear velocity	Equation 10
v	Flow velocity	Equation 1; Figure 12
w_s	Particle settling velocity	Equation 10
z	Altitude of channel bottom above an arbitrary datum	Equation 1; Figure 12
α	Velocity head coefficient	Equation 1; Figure 12
ϕ_1	Compass bearing of particle's long axis	Figure 58
ϕ_2	Inclination of upstream-dipping face of measured clast	Figure 58
γ	Specific weight of fluid	Equation 4
ρ	Fluid density	Equation 11
τ	Shear stress	Equation 4
τ_c	Critical shear stress for entrainment (for various transport modes)	Equation 12
ϖ	Stream power per unit area of bed	Equation 5

APPENDIX 2. EVIDENCE FOR MAXIMUM FLOOD STAGE

No.	USGS Quadrangle	Lat.	Long.	River Mile*	Minimum Altitude† (ft)	Maximum Altitude† (ft)	Minimum Altitude† (m)	Maximum Altitude† (m)	Method of Determining Altitude§	Upper/ Lower**	Feature	Source††
1	Asotin	46.344	117.038	146.0	1,020	1,060	310.9	323.1	Topography	L	Scoured basalt	
2	Lewiston Orchards S.	46.298	116.975	150.1	1,040	1,060	317.0	323.1	Survey	L	Eddy deposit	Webster and others (1982)
3	Captain John Rapids	46.132	116.935	164.0	1,080	1,120	329.2	341.4	Topography	L	Scoured basalt	
4	Limekiln Rapids	46.073	116.981	168.9	1,080	1,090	329.2	332.2	Survey	L	Crossed divide	
5	Limekiln Rapids	46.074	116.985	168.9	1,020	1,030	310.9	313.9	Survey	L	Eddy deposit	Webster and others (1982)
6	Limekiln Rapids	46.033	116.925	173.2	1,160	1,240	353.6	378.0	Topography	L	Eddy deposit	
7	Jim Creek Butte	45.993	116.907	176.5	1,320	1,360	402.3	414.5	Topography	L	Eddy deposit	Webster and others (1982)
8	Deadhorse Ridge	45.821	116.778	191.0	1,270	1,330	387.1	405.4	Altimeter	L	Eddy deposit	
9	Cactus Mountain	45.815	116.701	195.6	1,360	1,400	414.5	426.7	Topography	L	Eddy deposit	
10	Cactus Mountain	45.816	116.687	195.7	1,355	1,395	413.0	425.2	Survey	L	Eddy deposit	
11	Cactus Mountain	45.805	116.680	196.6	1,365	1,385	416.1	422.1	Survey	L	Eddy deposit	
12	Cactus Mountain	45.787	116.662	198.4	1,480	1,520	451.1	463.3	Topography	L	Eddy deposit	
13	Wolf Creek	45.767	116.586	203.2	1,480	1,520	451.1	463.3	Topography	L	Eddy deposit	
14	Wolf Creek	45.763	116.558	204.3	1,480	1,520	451.1	463.3	Topography	L	Eddy deposit	
15	Wolf Creek	45.760	116.553	204.6	1,440	1,520	438.9	463.3	Topography	L	Eddy deposit	
16	Kernan Point (15')	45.708	116.531	208.5	1,410	1,450	429.8	442.0	Topography	L	Eddy deposit	T. Vallier, USGS, Personal Communication (1989)
17	Grave Point	45.634	116.487	214.6	1,650	1,680	502.9	512.1	Survey	L	Eddy deposit	
18	Grave Point	45.645	116.461	214.8	1,655	1,675	504.4	510.5	Survey	L	Eddy deposit	
19	Kirkwood Creek	45.622	116.460	216.1	1,666	1,676	507.8	510.8	Survey	L	Scoured bedrock	
20	He Devil (15')	45.498	116.558	227.3	1,765	1,785	538.0	544.1	Survey	L	Eddy deposit	
21	He Devil (15')	45.393	116.639	236.1	1,670	1,750	509.0	533.4	Altimeter	L	Eddy deposit	
22	Cuprum (15')	45.245	116.700	247.6	1,735	1,735	528.8	528.8	?	L	Scoured bedrock	Stearns (1962)
23	Cuprum (15')	45.135	116.739	255.8	1,760	1,820	536.4	554.7	Topography	L	Expansion bar	
24	Homestead (15')	45.007	116.849	266.3	1,870	1,890	570.0	576.1	Survey	L	Scoured bedrock and rounded talus	

APPENDIX 2. **EVIDENCE FOR MAXIMUM FLOOD STAGE** (continued)

No.	USGS Quadrangle	Lat.	Long.	River Mile	Minimum Altitude (ft)	Maximum Altitude (ft)	Minimum Altitude (m)	Maximum Altitude (m)	Method of Determining Altitude	Upper/ Lower	Feature	Source
25	Oxbow	44.972	116.833	271.8	1,880	1,880	573.0	573.0	?	L	Scour features	Stearns (1962)
26	Oxbow	44.971	116.842	272.2	2,120	2,160	646.2	658.4	Topography	L	Scoured bedrock	
27	Brownlee Dam	44.837	116.898	283.5	2,090	2,090	637.0	637.0	?	L	Eddy deposit	Stearns (1962)
28	Brownlee Dam	44.810	116.914	285.6	2,190	2,210	667.5	673.6	Survey	L	Eddy deposit	
29	Sturgill Creek	44.659	117.103	302.0	2,261	2,271	689.2	692.2	Survey	L	Eddy deposit	
30	Sturgill Creek	44.636	117.113	303.8	2,280	2,300	694.9	701.0	Survey	L	Slackwater sediment	
31	Connor Creek	44.566	117.137	309.4	2,277	2,277	694.0	694.0	Spot Altitude	L	Scoured bedrock	
32	Connor Creek	44.538	117.186	314.3	2,280	2,320	694.9	707.1	Topography	L	Eddy deposit	
33	Connor Creek	44.517	117.175	314.5	2,328	2,328	709.6	709.6	Spot Altitude	L	Eddy deposit	
34	Olds Ferry NW	44.477	117.214	318.9	2,280	2,360	694.9	719.3	Topography	L	Eddy deposit	
35	Olds Ferry	44.368	117.233	327.2	2,353	2,373	717.2	723.3	Survey	L	Boulders	
36	Olds Ferry	44.316	117.206	332.3	2,360	2,400	719.3	731.5	Topography	L	Scoured bedrock	
37	Olds Ferry	44.321	117.232	333.3	2,411	2,431	734.9	741.0	Survey	L	Rounded pebbles	
38	Olds Ferry	44.264	117.146	339.1	2,437	2,457	742.8	748.9	Survey	L	Scoured bedrock, rounded pebbles	
39	Wilder	43.716	116.910	392.0	2,400	2,420	731.5	737.6	Topography	L	Slackwater deposits	Othberg and Fosberg (1989)
40	Givens Hot Springs	43.419	116.734	434.0	2,480	2,500	755.9	762.0	Topography	L	Gravel bar	
41	Walters Butte	43.357	116.590	442.0	2,492	2,496	759.6	760.8	Survey	L	Cobble deposit	
42	Walters Butte	43.301	116.511	447.9	2,527	2,540	770.2	774.2	Benchmark-Topo.	L	Stripped basalt, sand	Malde (1968)
43	Initial Point	43.285	116.296	450.3	2,570	2,610	783.3	795.5	Topography	L	Basalt cobbles	
44	Initial Point	43.290	116.447	452.2	2,580	2,620	786.4	798.6	Topography	L	Scoured bedrock	
45	Initial Point	43.270	116.396	455.5	2,528	2,538	770.5	773.6	Survey	L	Eddy deposit	
46	Sinker Butte	43.248	116.391	456.0	2,575	2,600	784.9	792.5	Topography	L	Eddy deposit	
47	Sinker Butte	43.221	116.376	457.6	2,525	2,550	769.6	777.2	Topography	L	Eddy deposit	
48	Wild Horse Butte	43.190	116.373	459.9	2,717	2,721	828.1	829.4	Survey	L	Gravel bar	Jarrett and Malde (1987)

APPENDIX 2. EVIDENCE FOR MAXIMUM FLOOD STAGE (continued)

No.	USGS Quadrangle	Lat.	Long.	River Mile	Minimum Altitude (ft)	Maximum Altitude (ft)	Minimum Altitude (m)	Maximum Altitude (m)	Method of Determining Altitude	Upper/ Lower	Feature	Source
49	Wild Horse Butte	43.190	116.373	459.9	2,750	2,756	838.2	840.0	Survey	L	Stripped basalt	Jarrett and Malde (1987)
50	Jackass Butte	43.085	116.198	474.0	2,850	2,875	868.7	876.3	Topography	U	Patterned ground	
51	Dorsey Butte	43.018	116.053	486.0	2,800	2,825	853.4	861.1	Topography	L	Stripped basalt	
52	C.J. Strike Dam	42.970	115.977	496.4	2,785	2,795	848.9	851.9	Survey	L	Eroded basalt	
53	C.J. Strike Dam	42.980	115.922	501.3	2,800	2,825	853.4	861.1	Topography	L	Eroded basalt, crossed divide	
54	C.J. Strike Dam	42.977	115.919	501.5	2,811	2,811	856.8	856.8	Spot Altitude	L	Crossed divide	
55	C.J. Strike Dam	42.975	115.916	501.7	2,825	2,850	861.1	868.7	Topography	U	Uncrossed divide	
56	Sand Dunes	42.960	115.654	519.0	2,865	2,875	873.3	876.3	Spot Alt./Topo.	L	Crossed divide, eroded basalt	
57	Indian Cove	42.952	115.527	525.5	2,850	2,875	868.7	876.3	Topography	L	Crossed divide, stripped basalt	
58	Pasadena Valley	42.940	115.148	552.4	2,875	2,900	876.3	883.9	Topography	U	Top of eroded scarp	
59	Pasadena Valley	42.921	115.148	553.5	2,849	2,849	868.4	868.4	Spot Altitude	L	Eddy deposit	
60	Pasadena Valley	42.922	115.141	553.5	2,900	2,925	883.9	891.5	Topography	L	Stripped basalt	
61	Pasadena Valley	42.908	115.425	555.2	2,836	2,840	864.4	865.6	Survey	L	Eddy deposits	
62	Pasadena Valley	42.912	115.137	555.2	2,925	2,950	891.5	899.2	Topography	L	Stripped basalt	
63	Ticeska	42.897	115.118	557.0	3,000	3,025	914.4	922.0	Topography	U	Top of eroded scarp	
64	Ticeska	42.910	115.123	557.3	2,992	2,992	912.0	912.0	Benchmark	L	Stripped basalt	
65	Ticeska	42.907	115.073	559.8	2,887	2,893	880.0	881.8	Survey	L	Eddy deposit	
66	Bliss	42.904	114.993	564.2	2,925	2,950	891.5	899.2	Topography	L	Eddy deposit	
67	Bliss	42.888	114.946	567.9	2,975	3,000	906.8	914.4	Topography	L	Eddy deposit	
68	Hagerman	42.819	114.893	574.3	3,012	3,012	918.1	918.1	Spot Altitude	L	Eddy deposit	
69	Hagerman	42.771	114.940	578.8	3,150	3,175	960.1	967.7	Topography	U	Top of eroded scarp	
70	Tuttle	42.781	114.874	581.3	3,070	3,080	935.7	938.8	Topography	L	Eddy deposit(?)	
71	Tuttle	42.768	114.850	583.9	3,126	3,126	952.8	952.8	Spot Altitude	L	Divide crossing(?)	
72	Thousand Springs	42.711	114.821	588.0	3,160	3,180	963.2	969.3	Topography	L	Stripped basalt	Malde (1968)

APPENDIX 2. EVIDENCE FOR MAXIMUM FLOOD STAGE (continued)

No.	USGS Quadrangle	Lat.	Long.	River Mile	Minimum Altitude (ft)	Maximum Altitude (ft)	Minimum Altitude (m)	Maximum Altitude (m)	Method of Determining Altitude	Upper/ Lower	Feature	Source
73	Thousand Springs	42.695	114.836	589.1	3,180	3,200	969.3	975.4	Topography	U	Top of eroded scarp	
74	Thousand Springs	42.673	114.797	592.0	3,200	3,210	975.4	978.4	Topography	L	Stripped basalt	
75	Thousand Springs	42.651	114.774	592.9	3,140	3,160	957.1	963.2	Topography	L	Slackwater deposits	
76	Niagara Springs	42.653	114.649	600.3	3,138	3,148	956.5	959.5	Survey	L	Eddy deposit	
77	Niagara Springs	42.655	114.627	601.3	3,430	3,440	1045.5	1048.5	Topography	L	Stripped basalt	
78	Jerome	42.653	114.615	602.0	3,451	3,451	1051.9	1051.9	Benchmark	L	Stripped basalt	
79	Jerome	42.640	114.563	605.1	3,490	3,500	1063.8	1066.8	Topography	L	Stripped basalt, gravel	
80	Jerome	42.642	114.553	605.7	3,510	3,530	1069.8	1075.9	Topography	L	Slackwater sediment	
81	Jerome	42.637	114.538	606.4	3,553	3,553	1083.0	1083.0	Benchmark	L	Stripped basalt, gravel	
82	Jerome	42.635	114.518	607.3	3,560	3,570	1085.1	1088.1	Topography	L	Stripped basalt	
83	Twin Falls	42.595	114.455	611.9	3,630	3,640	1106.4	1109.5	Topography	L	Stripped basalt	Malde (1968)
84	Twin Falls	42.583	114.400	614.7	3,750	3,760	1143.0	1146.0	Topography	L	Stripped basalt	
85	Kimberly	42.580	114.364	617.3	3,800	3,810	1158.2	1161.3	Topography	L	Stripped basalt	
86	Kimberly	42.577	114.364	617.5	3,841	3,841	1170.7	1170.7	Benchmark	U	Top of eroded scarp	
87	Kimberly	42.569	114.296	621.0	3,940	3,950	1200.9	1204.0	Topography	L	Stripped basalt	
88	Kimberly	42.565	114.283	621.9	3,980	4,000	1213.1	1219.2	Topography	L	Stripped basalt	
89	Eden	42.562	114.249	623.6	4,030	4,050	1228.3	1234.4	Topography	L	Stripped basalt	
90	Eden	42.542	114.189	627.0	4,060	4,070	1237.5	1240.5	Topography	L	Stripped basalt	
91	Eden	42.508	114.147	630.2	4,089	4,089	1246.3	1246.3	Spot Altitude	L	Base of eroded scarp	
92	Eden	42.509	114.127	632.0	4,130	4,140	1258.8	1261.9	Topography	U	Top of eroded scarp	
93	Murtaugh	42.499	114.127	632.0	4,098	4,098	1249.1	1249.1	Spot Altitude	L	Stripped basalt, boulders	
94	Murtaugh	42.511	114.118	633.0	4,115	4,115	1254.3	1254.3	Spot Altitude	L	Gravel	
95	Murtaugh	42.517	114.088	634.8	4,120	4,130	1255.8	1258.8	Topography	L	Boulder bar	

APPENDIX 2. EVIDENCE FOR MAXIMUM FLOOD STAGE (continued)

No.	USGS Quadrangle	Lat.	Long.	River Mile	Minimum Altitude (ft)	Maximum Altitude (ft)	Minimum Altitude (m)	Maximum Altitude (m)	Method of Determining Altitude	Upper/ Lower	Feature	Source
96	Murtaugh	42.520	114.083	635.0	4,170	4,180	1271.0	1274.1	Topography	U	Top of eroded scarp	
97	Murtaugh	42.551	114.047	637.0	4,140	4,150	1261.9	1264.9	Topography	L	Gravel in crossed divide	
98	Murtaugh	42.516	114.019	638.6	4,200	4,210	1280.2	1283.2	Topography	U	Top of eroded scarp	
99	Murtaugh	42.532	114.006	639.6	4,172	4,176	1271.6	1272.8	Survey	L	Boulder bar	
100	Burley SW	42.525	113.971	641.8	4,182	4,182	1274.7	1274.7	Benchmark	L	Stripped basalt	
101	Burley SW	42.537	113.946	643.4	4,255	4,265	1296.9	1300.0	Survey	L	Highest gravel	
102	Lake Walcott West	42.634	113.448	676.5	4,274	4,284	1302.7	1305.8	Survey	L	Highest gravel	
103	Lake Walcott West	42.632	113.410	679.0	4,260	4,270	1298.4	1301.5	Topography	L	Highest gravel	
104	Lake Walcott West	42.721	113.392	681.2	4,276	4,296	1303.3	1309.4	Survey	L	Highest gravel	
105	Lake Walcott SE	42.616	113.327	686.4	4,283	4,287	1305.5	1306.7	Survey	L	Highest gravel	
106	Lake Walcott SE	42.601	113.283	688.5	4,279	4,279	1304.2	1304.2	Benchmark	L	Gravel	
107	North Chapin Mountain	42.622	113.217	692.0	4,290	4,300	1307.6	1310.6	Topography	L	Eroded loess	
108	North Chapin Mountain	42.623	113.161	695.2	4,340	4,360	1322.8	1328.9	Topography	U	Top of eroded scarp	
109	Badger Peak	42.607	113.121	697.2	4,306	4,312	1312.5	1314.3	Survey	L	Slackwater sediment	
110	Register Rock	42.644	113.094	698.6	4,295	4,295	1309.1	1309.1	Spot Altitude	L	Eddy deposit	
111	Badger Peak	42.620	113.086	698.7	4,302	4,308	1311.2	1313.1	Survey	L	Slackwater sediment	
112	Register Rock	42.627	113.081	699.1	4,320	4,340	1316.7	1322.8	Topography	L	Slackwater sediment	
113	Register Rock	42.635	113.048	701.0	4,360	4,368	1328.9	1331.4	Survey	L	Slackwater sediment	
114	Neeley	42.653	112.988	704.0	4,420	4,460	1347.2	1359.4	Topography	L	Eddy deposit	
115	Neeley	42.683	112.967	706.9	4,480	4,500	1365.5	1371.6	Topography	L	Slackwater sediment	
116	Neeley	42.745	112.968	710.0	4,470	4,490	1362.5	1368.6	Topography	L	Thalweg of upland channel	
117	American Falls	42.802	112.754	720.0	4,470	4,490	1362.5	1368.6	Topography	L	Crossed divide; slackwater sediment	
118	Pocatello North	42.957	112.448	738.0	4,510	4,530	1374.6	1380.7	Topography	L	Slackwater sediment	
119	Pocatello North	42.957	112.446	738.0	4,530	4,540	1380.7	1383.8	Survey	U	Top of eroded scarp	

APPENDIX 2. EVIDENCE FOR MAXIMUM FLOOD STAGE (continued)

No.	USGS Quadrangle	Lat.	Long.	River Mile	Minimum Altitude (ft)	Maximum Altitude (ft)	Minimum Altitude (m)	Maximum Altitude (m)	Method of Determining Altitude	Upper/ Lower	Feature	Source
120	Pocatello North	42.921	112.451	741.0	4,548	4,558	1386.2	1389.3	Survey	U	Top of eroded scarp	
121	Pocatello North	42.879	112.478	743.0	4,578	4,588	1395.4	1398.4	Survey	U	Top of eroded scarp	
122	Pocatello South	42.851	112.447	745.6	4,621	4,625	1408.5	1409.7	Survey	U	Top of eroded scarp	
123	Pocatello South	42.831	112.393	748.6	4,658	4,668	1419.8	1422.8	Survey	U	Top of eroded scarp	
124	Pocatello South	42.812	112.401	749.2	4,647	4,667	1416.4	1422.5	Survey	U	Top of eroded scarp	
125	Inkom	42.805	112.362	750.5	4,689	4,699	1429.2	1432.3	Survey	U	Top of eroded scarp	
126	Inkom	42.801	112.341	753.0	4,811	4,831	1466.4	1472.5	Survey	L	Scoured bedrock	
127	McCammon	42.669	112.237	767.0	4,868	4,872	1483.8	1485.0	Survey	L	Eroded fan deposits	
128	McCammon	42.651	112.235	770.5	4,920	4,960	1499.6	1511.8	Topography	U	Top of eroded scarp	
129	Arimo	42.504	112.245	783.0	4,845	4,855	1476.8	1479.8	Survey	L	Slackwater sediment	
130	Downey West	42.417	112.214	788.0	4,854	4,864	1479.5	1482.5	Survey	L	Slackwater sediment	
131	Downey East	42.432	112.111	791.0	4,900	4,920	1493.5	1499.6	Topography	L	Eddy deposit	Bright and Ore (1987)
132	Downey East	42.390	112.119	792.8	4,980	5,020	1517.9	1530.1	Topography	U	Top of eroded scarp	
133	Downey East	42.399	112.080	794.1	5,040	5,060	1536.2	1542.3	Topography	U	Top of eroded scarp	
134	Oxford	42.358	112.046	797.2	5,088	5,088	1550.8	1550.8	Spot Altitude	L	Scoured bedrock	
135	Oxford	42.313	112.048	800.5	5,092	5,092	1552.0	1552.0	?	L	Bonneville shoreline	Bright and Ore (1987); Currey (1990)
	Eden Channel											
E1	Kimberly	42.598	114.306		3,830	3,840	1167.4	1170.4	Topography	L	Stripped basalt; gravel	
E2	Kimberly	42.606	114.284		3,880	3,890	1182.6	1185.7	Topography	U	Top of eroded scarp	
E3	Kimberly	42.616	114.264		3,900	3,910	1188.7	1191.8	Topography	L	Stripped basalt	

APPENDIX 2. EVIDENCE FOR MAXIMUM FLOOD STAGE (continued)

No.	USGS Quadrangle	Lat.	Long.	River Mile	Minimum Altitude (ft)	Maximum Altitude (ft)	Minimum Altitude (m)	Maximum Altitude (m)	Method of Determining Altitude	Upper/ Lower	Feature	Source
E4	Eden	42.617	114.229		3,950	3,960	1204.0	1207.0	Topography	U	Top of eroded scarp	
E5	Hunt	42.649	114.210		3,993	3,993	1217.1	1217.1	Spot Altitude	L	Stripped basalt	
E5	Hunt	42.652	114.179		4,010	4,030	1222.2	1228.3	Topography	L	Stripped basalt	
E6	Eden	42.602	114.170		4,050	4,060	1234.4	1237.5	Topography	U	Top of eroded scarp	
E7	Eden	42.616	114.170		4,030	4,050	1228.3	1234.4	Topography	L	Stripped basalt	
E8	Milner	42.612	114.081		4,080	4,090	1243.6	1246.6	Topography	L	Eroded basalt	
E9	Milner	42.584	114.039		4,120	4,130	1255.8	1258.8	Topography	L	Crossed divide	
E10	Milner	42.608	114.010		4,165	4,165	1269.5	1269.5	Spot Altitude	L	Eroded loess	
E11	Burley NW	42.634	113.986		4,180	4,190	1274.1	1277.1	Topography	L	Upland channel	
E12	Burley SW	42.614	113.932		4,252	4,256	1296.0	1297.2	Survey	L	Stripped basalt; pebbles	
E13	Burley SW	42.613	113.931		4,260	4,265	1298.4	1300.0	Topography	U	Patterned ground; uneroded loess	

*River Mile column indicates approximate position of maximum-stage evidence relative to the Snake River confluence with the Columbia River as labeled on USGS topographic quadrangles. For maps in which mileage is not indicated, the mileage was estimated. For locations within the Portneuf and Marsh Creek valleys, the distance upstream from the confluence with the Snake River was measured and added to the Snake River mileage at the Snake River/Portneuf River confluence.

†Altitudes of all features (except those identified by Stearns) were determined or redetermined in this study. The specified minimum altitude is the lower bound for the actual altitude of the feature considering the uncertainty in determining the feature's precise altitude; likewise, the maximum altitude is the upper bound of the feature's altitude. As discussed in the text, the estimated uncertainty depends on the method of survey or the contour interval of the local topographic map.

§In Method of Determining Altitude column, Topography = feature's altitude determined by location on USGS topographic maps; Survey = feature altitude surveyed (either by theodolite, or by tape and inclinometer) from a known altitude; Benchmark = feature at or at the same altitude as an established benchmark; Spot Altitude = feature at or at the same altitude as a labeled altitude on USGS topographic maps; Altimeter = altitude determined by altimeter.

**In Features column, features designated U are those interpreted as providing an upper limit on the maximum flood stage; that is, they are altitudes inferred not to have been flooded. Features designated L provide a lower limit on the maximum flood stage; that is, they are features interpreted to have been inundated at maximum flood stage.

††Sources refer to the reference where the feature was identified as a Bonneville Flood feature. All other features were identified in this study.

APPENDIX 3. SUMMARY OF BOULDER MEASUREMENTS AND HYDRAULIC CONDITIONS*

Hydraulic Parameters†						Twenty Largest Boulders§			Five Largest Boulders§		
Station	Velocity ($m \cdot sec^{-1}$)	Energy Slope	Depth (m)	Shear Stress ($N \cdot m^{-2}$)	Stream Power ($watts \cdot m^{-2}$)	Inter. Axis (cm)	Mass (tonnes)	Roundness	Inter. Axis (cm)	Mass (tonnes)	Roundness
Lewiston, Dug Bar, and Pittsburg Landing Reaches											
108A	3.7	0.000016	132.9	21	76	235	16.8	–	334	35.3	–
262	6.2	0.000046	127.9	64	404	–	–	–	292	–	–
99	5.6	0.000051	129.0	64	361	229	13.0	–	268	21.8	–
108B	2.7	0.00001	128.3	12	35	173	7.1	–	246	17.2	–
257A	5.2	0.000094	138.1	127	659	205	12.5	3.4	242	23.8	3.5
259	6.9	0.000067	129.6	84	584	174	7.9	3.0	236	16.2	3.3
261	5.6	0.000046	128.2	58	324	185	10.6	2.6	236	20.8	2.5
257B	6.6	0.000051	140.2	70	462	174	8.6	3.0	206	17.2	3.0
101	7.3	0.000056	152.9	84	608	145	4.2	–	204	8.8	–
102B	4.9	0.000026	156.6	40	196	414	4.8	–	202	11.9	–
257C	6.3	0.000049	143.3	69	432	152	6.1	2.8	194	10.6	3.0
260	7.8	0.000087	121.7	104	810	149	5.7	3.1	192	11.8	3.1
257D	6.1	0.000055	136.0	73	445	139	4.4	3.3	162	6.8	3.3
241	6.9	0.000114	86.4	97	668	136	3.3	3.6	155	4.7	3.7
97	12.6	0.000274	127.2	342	4,310	114	2.5	–	150	5.1	–
106B	3.1	0.000032	99.4	31	97	119	2.8	–	146	7.1	–
108F	2.4	0.000014	93.2	13	31	102	1.2	–	138	2.5	–
108C	2.7	0.000012	114.6	13	37	100	1.3	–	131	2.7	–
108G	2.0	0.000012	91.7	11	22	94	1.4	–	124	3.7	–
257E	6.4	0.000061	129.5	78	493	94	1.3	3.2	124	2.2	3.1
265	5.5	0.000034	145.0	48	264	107	1.8	3.1	124	3.0	3.2
108D	3.0	0.000016	106.9	17	51	93	1.3	–	120	2.8	–
108E	2.7	0.000014	103.8	14	39	90	1.1	–	118	2.3	–
102A	4.6	0.000026	142.9	37	167	85	0.77	–	116	1.6	–
263	3.9	0.000032	133.4	42	164	90	1.1	3.5	113	2.1	3.6
108H	1.6	0.000012	91.7	11	17	84	0.87	–	108	1.9	–
106A	3.1	0.000032	111.5	35	110	83	0.80	–	108	1.4	–
257F	6.1	0.000055	121.7	65	398	80	0.70	3.4	91	1.2	3.2
106C	4.2	0.000032	105.4	33	140	66	0.45	–	77	0.74	–

APPENDIX 3. SUMMARY OF BOULDER MEASUREMENTS AND HYDRAULIC CONDITIONS (continued)

Hydraulic Parameters						Twenty Largest Boulders			Five Largest Boulders		
Station	Velocity ($m \cdot sec^{-1}$)	Energy Slope	Depth (m)	Shear Stress ($N \cdot m^{-2}$)	Stream Power ($watts \cdot m^{-2}$)	Inter. Axis (cm)	Mass (tonnes)	Roundness	Inter. Axis (cm)	Mass (tonnes)	Roundness
Swan Falls Reach											
124	13.9	0.000436	89.2	381	5,284	530	205.2	3.5	624	336.4	3.6
114	14.3	0.000545	91.3	488	6,244	464	135.0	4.3	574	230.9	4.4
115	14.1	0.000501	93.2	458	5,762	442	134.3	3.9	534	238.3	3.8
299	11.6	0.000351	81.3	280	3,264	472	126.8	3.9	586	239.5	3.9
123	13.9	0.000440	86.0	371	5,165	426	104.8	3.9	500	179.0	3.9
116	12.7	0.000476	73.6	343	4,366	395	82.3	4.4	500	187.3	4.2
118	9.6	0.000384	56.4	212	2,032	343	55.7	4.2	452	101.6	4.1
280	7.6	0.000310	45.4	138	1,053	337	54.5	3.4	400	92.7	3.0
117	10.2	0.000380	59.4	221	2,259	338	50.8	4.2	404	95.1	3.5
291	11.7	0.000439	66.2	285	3,342	344	50.3	3.8	416	91.7	3.8
119	9.4	0.001154	51.0	577	5,449	295	32.8	4.0	352	59.6	4.1
120	8.5	0.000844	54.3	449	3,805	221	14.9	4.3	250	23.0	4.4
281	5.7	0.000199	35.4	69	395	219	14.4	4.3	260	25.6	4.2
292	8.9	0.000251	72.4	178	1,585	212	13.6	3.8	252	21.6	4.0
279	11.0	0.000520	58.1	296	3,261	204	13.1	4.2	246	24.0	4.0
293	11.1	0.000406	66.9	266	2,939	201	12.0	3.9	236	21.6	3.7
122	13.3	0.000496	75.9	369	4,809	173	8.1	4.3	194	12.8	4.3
273	4.8	0.000062	67.4	41	196	169	7.8	3.4	220	17.7	3.4
121	9.1	0.000790	63.3	490	4,454	179	7.3	4.4	206	11.5	4.2
282	5.1	0.000281	30.9	85	432	157	6.9	4.2	196	13.0	4.1
290	14.2	0.000551	80.6	435	6,186	165	6.8	4.2	202	10.2	3.9
272	5.2	0.000080	64.0	50	262	164	6.4	3.3	202	12.4	3.3
283	4.4	0.000219	28.9	62	274	107	1.5	4.4	130	2.4	4.5
289	3.8	0.000145	28.1	40	152	80	0.8	4.8	90	1.1	5.0
287	3.1	0.000095	27.8	26	80	49	0.2	4.7	61	0.35	4.4
285	2.6	0.000053	44.6	23	60	48	0.18	4.5	53	0.29	4.5
288	3.5	0.000108	29.2	31	109	50	0.17	5.2	56	0.26	5.3
286	3.3	0.000062	44.6	27	88	43	0.1	4.6	54	0.17	4.5
271	3.1	0.000146	46.3	66	202	31	0.04	6.8	38	0.066	3.5
284	2.5	0.000042	41.5	17	42	24	0.021	4.6	27	0.034	4.6
268	--	--	122.7	--	--	11	0.002	4.3	14	0.003	4.0
270	--	--	61.0	--	--	15	0.005	4.1	18	0.008	3.6

APPENDIX 3. SUMMARY OF BOULDER MEASUREMENTS AND HYDRAULIC CONDITIONS (continued)

Hydraulic Parameters						Twenty Largest Boulders			Five Largest Boulders		
Station	Velocity ($m \cdot sec^{-1}$)	Energy Slope	Depth (m)	Shear Stress ($N \cdot m^{-2}$)	Stream Power ($watts \cdot m^{-2}$)	Inter. Axis (cm)	Mass (tonnes)	Roundness	Inter. Axis (cm)	Mass (tonnes)	Roundness
Hagerman Reach											
194	13.1	0.000562	67.8	373	4,905	327	52.8	3.6	484	118.7	3.6
155	14.3	0.000698	65.4	448	6,399	312	47.8	4.0	462	108.7	3.6
154	15.1	0.000753	66.7	492	7,408	286	33.2	3.9	374	66.6	3.7
195	14.1	0.000623	71.0	433	6,129	294	41.7	3.2	372	78.0	3.0
203	3.2	0.000443	23.3	101	327	272	24.4	2.8	316	39.7	2.7
206	9.7	0.000245	81.8	197	1,899	267	22.1	3.9	314	34.2	3.7
166	9.8	0.001065	38.1	398	3,880	273	26.9	4.1	306	39.8	4.2
196	8.4	0.000175	69.8	120	1,011	232	19.4	3.9	296	42.2	3.8
204	4.2	0.000387	27.9	106	441	193	10.2	2.9	266	21.7	2.3
178	4.7	0.000173	34.0	58	272	192	12.0	3.7	258	29.3	3.6
167	11.8	0.001006	36.6	361	4,266	214	13.7	3.8	256	24.8	3.9
168	10.1	0.000756	34.3	254	2,563	208	11.4	3.7	248	18.9	3.7
205	4.0	0.000316	35.8	111	439	198	10.7	3.5	246	21.3	3.4
210	4.9	0.000082	67.2	54	263	202	12.6	3.8	246	22.5	3.8
207	6.5	0.000148	63.9	93	604	201	11.3	3.9	226	18.0	3.8
179	2.9	0.000082	26.3	21	62	182	8.1	4.1	224	15.5	4.0
197	10.5	0.000730	39.1	280	2,902	156	6.0	4.1	222	12.7	4.1
183	11.6	0.000661	53.7	348	4,041	176	8.2	3.8	220	12.4	3.8
198	4.1	0.000730	7.7	55	228	147	4.8	3.6	218	11.6	3.5
176	4.1	0.000173	30.9	52	217	169	6.8	3.9	212	13.7	3.8
199	4.7	0.000759	7.8	58	273	157	5.9	3.7	206	11.8	3.8
163	10.8	0.000662	43.4	281	3,045	171	7.3	4.0	206	14.2	3.6
190	12.5	0.000537	63.8	336	4,198	175	8.4	3.9	204	14.0	3.8
169	8.7	0.000595	31.7	185	1,601	163	5.5	3.8	198	9.3	3.5
185	13.3	0.000623	60.3	368	4,905	154	5.5	4.0	190	10.2	4.1
209	4.9	0.000082	61.1	49	239	153	5.7	4.0	190	10.8	3.9
189	12.0	0.000458	70.6	317	3,805	152	4.9	3.8	172	8.7	3.8
164	6.4	0.000257	42.3	107	682	136	3.8	4.1	168	7.1	3.6
184	12.6	0.000649	61.2	389	4,911	152	3.9	4.0	164	5.7	4.0
215	11.4	0.001625	33.4	532	6,060	129	2.8	3.6	158	4.5	3.4
180	2.7	0.000077	26.3	20	53	135	3.4	4.0	154	5.5	3.9
181	3.1	0.000103	25.8	26	80	120	2.5	4.2	140	3.8	4.2

APPENDIX 3. SUMMARY OF BOULDER MEASUREMENTS AND HYDRAULIC CONDITIONS (continued)

	Hydraulic Parameters					Twenty Largest Boulders			Five Largest Boulders		
Station	Velocity ($m \cdot sec^{-1}$)	Energy Slope	Depth (m)	Shear Stress ($N \cdot m^{-2}$)	Stream Power ($watts \cdot m^{-2}$)	Inter. Axis (cm)	Mass (tonnes)	Roundness	Inter. Axis (cm)	Mass (tonnes)	Roundness
172	7.9	0.000457	28.6	128	1,015	107	2.0	4.0	138	3.9	4.0
158	5.0	0.000168	36.4	60	298	117	2.1	4.4	136	3.8	4.2
211	3.6	0.000040	64.2	25	90	106	2.0	3.9	134	3.7	3.7
171	7.1	0.000457	27.1	121	865	115	2.5	4.2	132	4.1	4.0
170	7.1	0.000457	27.1	121	865	105	1.9	4.0	132	3.5	3.9
157	5.8	0.000144	41.2	58	337	109	1.7	–	128	2.6	–
177	5.9	0.000251	33.8	83	492	110	1.8	4.2	120	2.8	4.1
156	4.6	0.000123	41.2	50	230	81	0.70	4.2	93	1.3	4.1
214	3.5	0.000103	33.4	34	118	84	0.81	3.9	92	1.2	4.1
213	3.1	0.000082	33.9	27	86	69	0.50	4.3	76	0.8	3.9
212	2.9	0.000032	73.4	23	67	61	0.30	4.2	74	0.5	4.0
192	2.7	0.000059	32.6	19	52	53	0.20	4.2	60	0.3	4.4
182	3.8	0.000129	28.6	36	139	48	0.20	4.2	57	0.3	4.2
Murtaugh, Burley Basin, and Eden Reaches											
376	–	–	–	–	–	475	202	2.8	684	464	2.9
377	–	–	–	–	–	462	154	2.9	546	257	2.6
373	–	–	–	–	–	232	19	3.4	322	42.5	3.4
394	5.7	0.001083	25.2	267	1,518	216	14	3.3	250	23.4	3.4
396	4.1	0.000455	29.0	129	528	185	9.6	3.5	228	17.3	3.6
374	–	–	–	–	–	140	4.2	3.5	186	8.96	3.2
407	6.5	0.000959	27.9	262	1,704	121	2.6	3.7	142	4.55	3.7
406	9.2	0.004161	15.2	621	5,702	101	1.5	3.7	134	2.56	3.7
397	4.2	0.000734	19.6	141	598	32	0.041	3.9	49	0.09	4.0
364	7.7	0.000623	23.5	144	1,105	22	0.018	4.8	28	0.03	4.9
365	9.7	0.000467	13.4	61	182	22	0.018	4.0	26	0.029	2.8
383	3.6	0.000112	33.0	36	131	20	0.013	3.7	25	0.026	3.1
370	4.1	0.000218	24.9	53	218	17	0.008	3.6	20	0.016	3.7
382B	4.6	0.000161	35.7	56	259	14	0.004	3.6	17	0.008	3.4
Pocatello Reach											
415	10.0	0.000460	52.3	236	2,358	110	1.7	3.3	144	3.0	3.2
422	2.7	0.000043	40.9	17	47	64	0.32	3.4	95	0.70	3.2
428	2.8	0.000032	64.8	21	56	51	0.21	3.5	73	0.50	3.4
427	3.5	0.000044	61.5	27	94	40	0.096	3.6	49	0.18	3.3

APPENDIX 3. SUMMARY OF BOULDER MEASUREMENTS AND HYDRAULIC CONDITIONS (continued)

	Hydraulic Parameters					Twenty Largest Boulders			Five Largest Boulders		
Station	Velocity ($m \cdot sec^{-1}$)	Energy Slope	Depth (m)	Shear Stress ($N \cdot m^{-2}$)	Stream Power ($watts \cdot m^{-2}$)	Inter. Axis (cm)	Mass (tonnes)	Roundness	Inter. Axis (cm)	Mass (tonnes)	Roundness
442	2.7	0.000064	47.5	30	81	32	0.053	3.8	38	0.083	3.8
421	2.9	0.000055	37.7	21	58	27	0.037	3.7	38	0.089	3.9
420	4.0	0.000075	42.0	31	124	27	0.034	3.6	33	0.064	3.5
418	3.2	0.000043	48.6	21	64	20	0.012	3.7	25	0.022	3.6
424	2.1	0.000055	25.5	13	30	17	0.009	3.7	23	0.016	3.6
426	2.3	0.000032	43.5	13	32	16	0.006	3.7	18	0.011	3.7
425	2.9	0.000059	34.3	20	56	15	0.005	3.8	18	0.009	4.0
432	4.5	0.000061	82.5	49	220	13	0.004	3.6	17	0.006	3.5
419	2.3	0.000065	55.8	36	82	13	0.004	4.0	15	0.009	4.1

*Complete boulder data available from author.

†Hydraulic parameters are from the results of the step-backwater flow reconstructions as described in Figure 53. The discharges for the flow reconstructions at each reach are listed in Table 1.

§Boulder measurement methods and site selection are described in the text and in Figure 58.

REFERENCES CITED

Allen, J.R.L., 1968, Current ripples, their relation to patterns of water and sediment motion: Amsterdam, North Holland, 433 p.

Bagnold, R. A., 1966, An approach to the sediment transport problem from general physics: U.S. Geological Survey Professional Paper 422-I, 37 p.

—— , 1977, Bed load transport by natural rivers: Water Resources Research, v. 13, p. 303–312.

—— , 1980, An empirical correlation of bedload transport rates in flumes and natural rivers: Proceedings of the Royal Society, v. 372-A, p. 453–473.

Baker, V. R., 1973, Paleohydrology and sedimentology of Lake Missoula flooding in eastern Washington: Geological Society of America Special Paper 144, 79 p.

—— , 1983, Late Pleistocene fluvial systems, *in* Porter, S. C., ed., The late Pleistocene: Minneapolis, University of Minnesota Press, p. 115–129.

—— , 1984, Flood sedimentation in bedrock fluvial systems, *in* Koster, E. H., and Steel, R. J., eds., Sedimentology of gravels and conglomerates: Canadian Society of Petroleum Geologists Memoir 10, p. 87–98.

Baker, V. R., and Bunker, R. C., 1985, Cataclysmic late Pleistocene flooding from glacial Lake Missoula; A review: Quaternary Science Reviews, v. 4, p. 1–41.

Baker, V. R., and Costa, J. E., 1987, Flood power, *in* Mayer, L., and Nash, D., eds., Catastrophic flooding: London, Allen and Unwin, p. 1–24.

Baker, V. R., and Komar, P. D., 1987, Cataclysmic flood processes and landforms, *in* Baker, V. R., Greeley, R., Komar, P. D., Swanson, D. A., and Waitt, R. B., Jr., The Columbia and Snake River Plains, *in* Graf, W. L., ed., Geomorphic systems of North America: Boulder, Colorado, Geological Society of America, The Geology of North America, Centennial Special Volume 2, p. 423–443.

Baker, V. R., and Ritter, D. F., 1975, Competence of rivers to transport coarse bed load material: Geological Society of America Bulletin, v. 86, p. 975–978.

Barnes, H. L., 1956, Cavitation as a geological agent: American Journal of Science, v. 254, p. 493–505.

Bedient, P. P., and Huber, W. C., Hydrology and floodplain analysis: Reading, Massachusetts, Addison-Wesley, 650 p.

Benson, L. V., and Thompson, R. S., 1987, The physical record of lakes in the Great Basin, *in* Ruddiman, W. F., and Wright, H. E., eds., North America and adjacent oceans during the last deglaciation: Geological Society of America, The Geology of North America, Centennial Special Volume K-3, p. 241–260.

Blatt, H., Middleton, G. V., and Murray, R. C., 1980, Origin of sedimentary rocks: Englewood Cliffs, New Jersey, Prentice-Hall, 782 p.

Bretz, J H., 1923, The Channeled Scabland of the Columbia Plateau: Journal of Geology, v. 31, p. 617–649.

—— , 1929, Valley deposits immediately east of the Channeled Scabland of Washington: Journal of Geology, v. 37, p. 393–427, 505–541.

Bretz, J H., Smith, H.T.U., and Neff, G. E., 1956, Channeled Scabland of Washington; New data and interpretations: Geological Society of America Bulletin, v. 67, p. 957–1,049.

Bright, R. C., and Ore, H. T., 1987, Evidence for the spillover of Lake Bonneville, southeastern Idaho, *in* Bues, S. S., ed., Geological Society of America Centennial Field Guide, Rocky Mountain Section, p. 143–146.

Carr, W. D., and Trimble, D. E., 1963, Geology of the American Falls Quadrangle, Idaho: U.S. Geological Survey Bulletin 1121-G, 44 p.

Chow, V. T., 1959, Open-channel hydraulics: New York, McGraw-Hill, 690 p.

Church, M., 1978, Paleohydrological reconstructions from a Holocene valley fill, *in* Miall, A. D., ed., Fluvial sedimentology: Canadian Society of Petroleum Geologists Memoir 5, p. 743–772.

Costa, J. E., 1983, Paleohydraulic reconstruction of flash-flood peaks from

boulder deposits in the Colorado Front Range: Geological Society of America Bulletin, v. 94, p. 986–1,004.
——, 1987, Hydraulics and basin morphology of the largest flash floods in the conterminous United States: Journal of Hydrology, v. 93, p. 313–338.
Covington, J. R., Whitehead, R. L., and Weaver, J. N., 1985, Ancestral canyons of the Snake River: Geology and hydrology of canyon-fill deposits in the Thousand Springs area, south-central Snake River Plain, Idaho: Geological Society of America, Rocky Mountain Section, 38th Annual Meeting, April 22–24, 1985, Boise, Idaho, Guidebook, 30 p.
Craig, R. G., 1987, Dynamics of a Missoula flood, *in* Mayer, L., and Nash, D., eds., Catastrophic flooding: London, Allen and Unwin, p. 305–332.
Currey, D. R., 1990, Quaternary palaeolakes in the evolution of semidesert basins, with special emphasis on Lake Bonneville and the Great Basin, U.S.A.: Palaeogeography, Palaeoclimatology, Palaeoecology, v. 76, p. 189–214.
Currey, D. R., and Oviatt, C. G., 1985, Durations, average rates, and probable causes of Lake Bonneville expansions, stillstands, and contractions during the last deep-lake cycle, 32,000 to 10,000 years ago: Geography Journal of Korea, v. 10, p. 1085–1099.
Currey, D. R., Atwood, G., and Mabey, D. R., 1983a, Major levels of Great Salt Lake and Lake Bonneville: Salt Lake City, Utah, Utah Geological and Mineral Survey Map 73.
Currey, D. R., Oviatt, C. G., and Plyler, G. B., 1983b, Lake Bonneville stratigraphy, geomorphology, and isostatic deformation in west-central Utah, *in* Geologic excursions in neotectonics and engineering geology in Utah: Utah Geological and Mineral Survey Special Study 62, p. 63–82.
Dunne, T., 1980, Formation and controls of channel networks: Progress in Physical Geography, v. 4, p. 211–239.
Einstein, H. A., Anderson, A. G., and Johnson, J. W., 1940, A distinction between bed load and suspended load in natural streams: American Geophysical Union Transactions, v. 19, p. 628–632.
Elfström, Åsa, 1983, The Bldakatj boulder delta, Lapland, northern Sweden: Geografiska Annaler, v. 65A, p. 201–225.
——, 1987, Large boulder deposits and catastrophic floods: Geografiska Annaler, v. 69A, p. 101–121.
Ely, L. L., and Baker, V. R., 1985, Reconstructing paleoflood hydrology with slackwater deposits, Verde River, Arizona: Physical Geography, v. 5, p. 103–126.
Fahnestock, R. K., 1963, Morphology and hydrology of a glacial stream—White River, Mount Rainier, Washington: U.S. Geological Survey Professional Paper 442-A, p. 1–70.
Feldman, A. D., 1981, HEC models for water resources system simulation: Theory and experience: Advances in Hydroscience, v. 12, p. 297–423.
Fitzgerald, J. F., 1982, Geology and basalt stratigraphy of the Weiser embayment, west-central Idaho, *in* Bonnichsen, B., and Breckinridge, R. M., eds., Cenozoic geology of Idaho: Idaho Bureau of Mines and Geology Bulletin 26, p. 103–128.
Flint, R. F., 1938, Origin of the Cheney-Palouse scabland tract: Geological Society of America Bulletin, v. 49, p. 461–524.
Florsheim, J. L., and Keller, E. A., 1989, Hydrologic geometry of riffle-pool sequences: A modeling approach: Geological Society of America Abstracts with Programs, v. 21, no. 5, p. 79.
Fread, D. L., 1988, The National Weather Service DAMBRK model: Theoretical background/user documentation: Silver Spring, Maryland, National Weather Service, 123 p.
Gilbert, G. K., 1878, The ancient outlet of Great Salt Lake: American Journal of Science (3rd ser.), v. 15, p. 256–259.
——, 1890, Lake Bonneville: U.S. Geological Survey Monograph 1, 438 p.
——, 1914, The transportation of debris by running water: U.S. Geological Survey Professional Paper 86, 263 p.
Glancy, P. A., and Harmsen, L., 1975, A hydrologic assessment of the September 14, 1974, flood in Eldorado Canyon, Nevada, U.S. Geological Survey Professional Paper 930, 28 p.
Grimm, C. I., and Leupold, N., 1939, Hydraulic data pertaining to the design of rock revetment, U.S. Engineer Office, North Pacific Division, Portland, Oregon, 32 p. (unpublished report).
Helley, E. J., 1969, Field measurement of the initiation of large bed particle motion in Blue Creek near Klamath, California: U.S. Geological Survey Professional Paper 562-G, 19 p.
Hjülstrom, F., 1935, Studies in the morphological activity of rivers as illustrated by the River Fyris: Bulletin of the Geological Institute University of Uppsala, v. 25, p. 221–527.
Hunt, C. B., ed., 1982, Pleistocene Lake Bonneville, ancestral Great Salt Lake, as described in the notebooks in G. K. Gilbert, 1875–1880: Brigham Young University Geology Studies, v. 29, pt. 1, 225 p.
Hydrologic Engineering Center, 1985, HEC-2 water surface profiles users manual: Davis, California, Hydrologic Engineering Center (U.S. Army Corps of Engineers), 37 p.
Inbar, M., and Schick, A. P., 1979, Bedload transport associated with high stream power, Jordan River, Israel: Proceedings of the National Academy of Science, v. 76, p. 2515–2517.
Jarrett, R. D., 1984, Hydraulics of high-gradient streams: Journal of the Hydraulics Division of the American Society of Civil Engineers, v. 110, p. 1519–1539.
——, 1987, Errors in slope-area computations of peak discharges in mountain streams: Journal of Hydrology, v. 96, p. 53–67.
Jarrett, R. D., and Malde, H. E., 1987, Paleodischarge of the late Pleistocene Bonneville Flood, Snake River, Idaho, computed from new evidence: Geological Society of America Bulletin, v. 99, p. 127–134.
Johansson, C. E., 1976, Structural studies of frictional sediments: Geografiska Annaler, v. 58A, p. 201–300.
Jopling, A. V., 1963, Hydraulic studies on the origin of bedding: Sedimentology, v. 2, p. 115–121.
Jopling, A. V., and Walker, R. G., 1968, Morphology and origin of ripple-drift cross-lamination, with examples from the Pleistocene of Massachusetts: Journal of Sedimentary Petrology, v. 38, p. 971–984.
Kehew, A. E., 1977, Environmental geology of Lewiston, Idaho, and vicinity [Ph.D. thesis]: Moscow, University of Idaho, 211 p.
Kellerhals, R., 1967, Stable channels with gravel-paved beds: Journal of the Waterways and Harbors Division, American Society of Civil Engineers, v. 93, WW1, p. 63–84.
Komar, P. D., 1979, Comparisons of the hydraulics of water flows in Martian outflow channels with flows of similar scale on Earth: Icarus, v. 37, p. 156–181.
——, 1980, Modes of sediment transport in channelized water flows with ramifications to the erosion of the Martian outflow channels: Icarus, v. 42, p. 317–329.
——, 1987, Selective gravel entrainment and the empirical evaluation of flow competence: Sedimentology, v. 34, p. 1165–1176.
Komar, P. D., and Carling, P. A., 1991, Grain sorting in gravel-bed streams and the choice of particle sizes for flow competence evaluations: Sedimentology, v. 38, p. 489–502.
Komar, P. D., and Li, Z., 1988, Applications of grain pivoting and sliding analyses to selective entrainment of gravel and to flow competence evaluations: Sedimentology, v. 35, p. 681–695.
Laity, J. E., and Malin, M. C., 1985, Sapping processes and development of theater-headed valley networks on the Colorado Plateau: Geological Society of America Bulletin, v. 95, p. 203–217.
Lane, E. W., and Carlson, E. J., 1953, Some factors affecting the stability of canals constructed in coarse granular materials: Proceedings of the International Association of Hydrologic Research, p. 37–48.
Leopold, L. P., Wolman, M. G., and Miller, J. P., 1964, Fluvial processes in geomorphology: San Francisco, California, Freeman, 522 p.
Lewis, G. C., and Fosberg, M. A., 1982, Distribution and character of loess and loess soils in southeastern Idaho, *in* Bonnichsen, B., and Breckenridge, R. M., eds., Cenozoic geology of Idaho: Idaho Bureau of Mines and Geology Bulletin 26, p. 705–716.
Mackin, J. H., 1963, Rational and empirical methods of investigation in geology,

in Albritton, C. C., ed., The fabric of geology: Reading, Massachusetts, Addison-Wesley, p. 135–163.

Malde, H. E., 1960, Evidence in the Snake River Plain, Idaho, of a catastrophic flood from Pleistocene Lake Bonneville, *in* Short papers in the geological sciences: U.S. Geological Survey Professional Paper 400-B, p. B295–B297.

—— , 1968, The catastrophic late Pleistocene Bonneville Flood in the Snake River Plain, Idaho: U.S. Geological Survey Professional Paper 596, 52 p.

—— , 1971, History of the Snake River canyon indicated by revised stratigraphy of Snake River Group near Hagerman and King Hill, Idaho, with a section on paleomagnetism by Allan Cox: U.S. Geological Survey Professional Paper 644-F, 21 p.

—— , 1982, The Yahoo Clay, a lacustrine unit impounded by the McKinney Basalt in the Snake River canyon near Bliss, Idaho, *in* Bonnichsen, B., and Breckenridge, R. M., eds., Cenozoic geology of Idaho: Idaho Bureau of Mines and Geology Bulletin 26, p. 617–628.

—— , 1987, Shoshone Falls, Idaho; A Pleistocene relic of the catastrophic Bonneville Flood, *in* Bues, S. S., ed., Geological Society of America Centennial Field Guide, Rocky Mountain Section, p. 143–146.

—— , 1989a, Geologic map of the Bruneau Formation in the Walters Butte and Initial Point Quadrangles, southwestern Idaho: U.S. Geological Survey Miscellaneous Field Studies Maps MF-2063-A, scale 1:24,000.

—— , 1989b, Geologic map of the Bruneau Formation in the Sinker Butte and Wild Horse Butte Quadrangles, southwestern Idaho: U.S. Geological Survey Miscellaneous Field Studies Maps MF-2063-B, scale 1:24,000.

—— , 1989c, Geologic map of the Bruneau Formation in the Oreana and Castle Butte Quadrangles, southwestern Idaho: U.S. Geological Survey Miscellaneous Field Studies Maps MF-2063-C, scale 1:24,000.

—— , 1989d, Geologic map of the Bruneau Formation in the Jackass Butte and Dorsey Butte Quadrangles, southwestern Idaho: U.S. Geological Survey Miscellaneous Field Studies Maps MF-2063-D, scale 1:24,000.

—— , 1989e, Geologic map of the Bruneau Formation in the C. J. Strike Dam and Bruneau Quadrangles, southwestern Idaho: U.S. Geological Survey Miscellaneous Field Studies Maps MF-2063-E, scale 1:24,000.

—— , 1991, Quaternary geology and structural history of the Snake River Plain, Idaho and Oregon, *in* Morrison, R. B., ed., Quaternary nonglacial geology, conterminous United States: Boulder, Colorado, Geological Society of America, The Geology of North America, v. K-2, p. 251–281.

Malde, H. E., and Powers, H. A., 1962, Upper Cenozoic stratigraphy of western Snake River Plain, Idaho: Geological Society of America Bulletin, v. 73, p. 1,197–1,220.

—— , 1972, Geologic map of the Glenns Ferry–Hagerman area, west-central Snake River Plain, Idaho: U.S. Geological Survey Miscellaneous Geologic Investigations Map I-696, scale 1:48,000, 2 sheets.

Malde, H. E., Powers, H. A., and Marshall, C. H., 1963, Reconnaissance geologic map of west-central Snake River Plain, Idaho: U.S. Geological Survey Miscellaneous Geologic Investigations Map I-373, scale 1:125,000.

Matthes, G. H., 1947, Macroturbulence in natural stream flow: EOS American Geophysical Union Transactions, v. 28, p. 255–262.

Middleton, G. V., and Southard, J. B., 1984, Mechanics of sediment movement: Lecture Notes for Society of Economic Paleontologists and Mineralogists Short Course No. 3, 401 p.

Milhous, R. T., 1973, Sediment transport in a gravel-bottomed stream [Ph.D. thesis]: Corvallis, Oregon State University, 232 p.

Mills, H. H., 1979, Downstream rounding of pebbles—A quantitative review: Journal of Sedimentary Petrology, v. 49, p. 295–302.

O'Brien, M. P., and Rindlaub, B. D., 1934, The transportation of bedload by streams: Transactions, American Geophysical Union, v. 15, p. 593–603.

O'Connor, J. E., and Baker, V. R., 1992, Peak discharges from Glacial Lake Missoula: Geological Society of America Bulletin, v. 104, p. 267–279.

O'Connor, J. E., and Webb, R. H., 1988, Hydraulic modeling for paleoflood analysis, *in* Baker, V. R., Kochel, R. C., and Patton, P. C., eds., Flood geomorphology: New York, Wiley, p. 403–418.

O'Connor, J. E., Webb, R. H., and Baker, V. R., 1986, Paleohydrology of pool-and-riffle pattern development, Boulder Creek, Utah: Geological Society of America Bulletin, v. 97, p. 410–420.

Othberg, K. L., and Fosberg, M. A., 1989, Late Pleistocene to late Pliocene soils on terraces of the Boise River, western Snake River Plain, Idaho: Geological Society of America Abstracts with Programs, v. 21, no. 5, p. 127.

Partridge, J. B., and Baker, V. R., 1987, Palaeoflood hydrology of the Salt River, Arizona: Earth Surface Processes and Landforms, v. 12, p. 109–125.

Pettijohn, F. J., Potter, P. E., and Siever, R., 1972, Sand and sandstone: New York, Springer-Verlag, 618 p.

Pierce, K. L., Fosberg, M. A., Scott, W. E., Lewis, G. C., and Colman, S. M., 1982, Loess deposits of southeastern Idaho: Age and correlation of the upper two loess units, *in* Bonnichsen, B., and Breckenridge, R. M., eds., Cenozoic geology of Idaho: Idaho Bureau of Mines and Geology Bulletin 26, p. 717–725.

Powers, M. C., 1953, A new roundness scale for sedimentary particles: Journal of Sedimentary Petrology, v. 23, p. 117–119.

Raudkivi, A. J., 1976, Loose boundary hydraulics (2nd edition): Oxford, Pergamon Press, 397 p.

Reidel, S. P., 1982, Stratigraphy of the Grande Ronde basalt, Columbia River Basalt Group, from the lower Salmon River and northern Hells Canyon area, Idaho, Oregon, and Washington, *in* Bonnichsen, B., and Breckenridge, R. M., eds., Cenozoic geology of Idaho: Idaho Bureau of Mines and Geology Bulletin 26, p. 77–101.

Ritter, J. R., 1967, Bed-material movement, Middle Fork Eel River, California: U.S. Geological Survey Professional Paper 575-C, p. 219–221.

Rouse, H., 1946, Elementary mechanics of fluids: New York, Wiley, 376 p.

Rubey, W. W., 1938, The force required to move particles on a stream bed: U.S. Geological Survey Professional Paper 189, p. 121–141.

Russell, I. C., 1902, Geology and water resources of the Snake River Plains of Idaho: U.S. Geological Survey Bulletin 199, 192 p.

Schmidt, J. C., 1990, Recirculating flow and sedimentation in the Colorado River in Grand Canyon, Arizona: Journal of Geology, v. 98, p. 709–724.

Scott, K. M., and Gravlee, G. C., Jr., 1968, Flood surge on the Rubicon River, California—Hydrology, hydraulics and boulder transport: U.S. Geological Survey Professional Paper 422-M, 40 p.

Scott, W. E., 1982, Surficial geologic map of the eastern Snake River Plain and adjacent areas, 111° to 115°W., Idaho and Wyoming: U.S. Geological Survey Miscellaneous Investigations Series Map I-1372, scale 1:250,000, 2 sheets.

Scott, W. E., Pierce, K. L., Bradbury, J. P., and Forester, R. M., 1982, Revised Quaternary stratigraphy and chronology in the American Falls area, southeastern Idaho, *in* Bonnichsen, B., and Breckenridge, R. M., eds., Cenozoic geology of Idaho: Idaho Bureau of Mines and Geology Bulletin 26, p. 581–595.

Scott, W. E., McCoy, W. D., Shroba, R. R., and Rubin, M., 1983, Reinterpretation of the exposed record of the last two cycles of Lake Bonneville, western United States: Quaternary Research, v. 20, p. 261–285.

Sedimentary Petrology Seminar, 1965, Gravel fabric in Wolf Run: Sedimentology, v. 4, p. 273–283.

Shepherd, R. G., and Schumm, S. A., 1974, An experimental study of river incision: Geological Society of America Bulletin, v. 85, p. 257–268.

Shroba, R. R., Schmidt, P. W., Crosby, E. J., Hansen, W. R., and Soule, J. M., 1979, Storm and flood of July 31-August 1, 1976, in the Big Thompson River and Cache la Poudre River Basins, Larimer and Weld Counties, Colorado. Part B, Geologic and geomorphic effects in the Big Thompson Canyon area, Larimer County: U.S. Geological Survey Professional Paper 1115, p. 87–152.

Shroder, J. F., and Lowndes, T., 1989, Threshold modeling, slope failure and the Bonneville Flood: Geological Society of America Abstracts with Programs, v. 21, no. 6, p. 60.

Smith, G. A., 1992, Missoula Flood dynamics and magnitude inferred from sedimentology of slackwater deposits on the Columbia Plateau, Washington: Geological Society of America Bulletin (in press).

Sneed, E. D., and Folk, R. L., 1958, Pebbles in the lower Colorado River, Texas; A study in particle morphogenesis: Journal of Geology, v. 66, p. 114–150.

Stearns, H. T., 1962, Evidence of Lake Bonneville flood along Snake River below King Hill, Idaho: Geological Society of America Bulletin, v. 73, p. 385–387.

Stearns, H. T., and Anderson, A. L., 1966, Geology of the oxbow on Snake River near Homestead, Oregon: Idaho Bureau of Mines and Geology, Pamphlet 136, 23 p.

Stearns, H. T., Crandall, L., and Steward, W. G., 1938, geology and ground-water resources of the Snake River Plain in southeastern Idaho: U.S. Geological Survey Water-Supply Paper 774, 268 p.

Sundborg, A., 1956, The River Klaralven; A study of fluvial processes: Geografiska Annaler, v. 38, p. 125–316.

Trimble, D. E., 1976, Geology of the Michaud and Pocatello Quadrangles, Bannock and Power Counties, Idaho: U.S. Geological Survey Bulletin 1400, 88 p.

Trimble, D. E., and Carr, W. J., 1961a, The Michaud delta and Bonneville River near Pocatello, Idaho: U.S. Geological Survey Professional Paper 424-B, p. B164–B166.

——, 1961b, Late Quaternary history of the Snake River in the American Falls region, Idaho: Geological Society of America Bulletin, v. 72, p. 1,739–1,748.

Vallier, T. L., 1977, The Permian and Triassic Seven Devils Group, western Idaho and northeastern Oregon: U.S. Geological Survey Bulletin 1437, 58 p.

Waitt, R. B., 1980, About forty last-glacial Lake Missoula jökulhlaups from glacial Lake Missoula: Geological Society of America Bulletin, v. 95, p. 1,271–1,286.

——, 1985, Case for periodic, colossal jökulhlaups from glacial Lake Missoula: Geological Society of America Bulletin, v. 95, p. 1271–1286.

Webb, R. H., 1985, Late Holocene flooding on the Escalante River, south-central Utah [Ph.D. thesis]: Tucson, University of Arizona, 204 p.

Webb, R. H., O'Connor, J. E., and Baker, V. R., 1988, Paleohydrologic reconstruction of flood frequency on the Escalante River, south-central Utah, *in* Baker, V. R., Kochel, R. C., and Patton, P. C., eds., Flood geomorphology: New York, Wiley, p. 403–418.

Webster, G. D., Kuhns, M.J.P., and Waggoner, G. L., 1982, Late Cenozoic gravels in Hells Canyon and the Lewiston basin, Washington and Idaho, *in* Bonnichsen, B., and Breckenridge, R. M., eds., Cenozoic geology of Idaho: Idaho Bureau of Mines and Geology Bulletin 26, p. 669–683.

Wheeler, H. E., and Cook, R. F., 1954, Structural and stratigraphic significance of the Snake River capture, Idaho-Oregon: Journal of Geology, v. 62, p. 525–536.

Williams, G. P., 1983, Paleohydrological methods and some examples from Swedish fluvial environments, I—Cobble and boulder deposits: Geografiska Annaler, v. 65A, p. 227–243.

Williams, J. S., and Milligan, J. H., 1968, Bedrock configuration and altitude, Red Rock Pass, outlet of Lake Bonneville, southeastern Idaho: University of Wyoming Contributions to Geology, v. 7, no. 1, p. 67–72.

Wohl, E. E., 1992a, Gradient irregularity in the Herbert Gorge of northeastern Australia: Earth Surface Processes and Landforms, v. 17, p. 69–84.

——, 1992b, Bedrock benches and boulder bars: Floods in the Burdekin Gorge of Australia: Geological Society of America Bulletin, v. 104, p. 770–778.

Wolman, M. G., and Eiler, J. P., 1958, Reconnaissance study of erosion and deposition produced by the flood of August 1955 in Connecticut: American Geophysical Union Transactions, v. 39, p. 1–14.

MANUSCRIPT ACCEPTED BY THE SOCIETY MAY 20, 1992

Printed in U.S.A.

Typeset by WESType Publishing Services, Inc., Boulder, Colorado
Printed in U.S.A. by Malloy Lithographing, Inc., Ann Arbor, Michigan